Science education and development: planning
and policy issues at secondary level

Science education
and development:

planning and policy issues
at secondary level

by
Françoise Caillods
Gabriele Göttelmann-Duret
Keith Lewin

International Institute for Educational Planning

PERGAMON

An Imprint of Elsevier Science

The designations employed and the presentation of material throughout this review do not imply the expression of any opinion whatsoever on the part of UNESCO or IIEP concerning the legal status of any country, territory, city or area or its authorities, or concerning its frontiers or boundaries.

The IIEP is financed by UNESCO and by voluntary contributions from Member States. In recent years the following Member States have provided voluntary contributions to the Institute: Denmark, Germany, Iceland, India, Ireland, Norway, Sweden, Switzerland and Venezuela.

Published by the
United Nations Educational, Scientific and Cultural Organization,
7 place de Fontenoy, F 75352 Paris 07 SP, France, and
Elsevier Science Ltd,
The Boulevard, Langford Lane, Kidlington, Oxford, OX5 1GB, UK
Printed in France by Imprimerie STEDI, 75018 Paris

ISBN 92-803-1160-3 (limp-bound edition)
ISBN 0-08-0427898 (hard-back edition)
© UNESCO 1997   IIEP/ko'f

Preface

The guiding principle of the International Institute for Educational Planning (IIEP) is the provision of assistance to governments, especially of developing nations, to adapt their educational policies to the changes taking place in modern society. We know that investment in education contributes significantly to economic and social development. Science and modern technology are vital to future growth, to production, agricultural progress, and the development of new services. Thus, human resource development in up-to-date knowledge, innovative aptitudes, and technical know-how is indispensable to the sustained development of nations. Science education, in particular, plays a decisive role: it forms a launching pad from which young people prepare themselves for the future.

It is over 30 years since the first large-scale investments in science education were made throughout the developing world. Since then, investment in science education has appeared as a priority area in almost every national development plan and policy statement. Most countries have undergone cycles of curriculum development; much time and money have been invested in training science teachers, and expensive facilities and equipment have been purchased to support discovery-oriented science training.

With the exception of a few remarkable cases, in spite of these efforts, shortages in the supply of science-trained teachers persist, achievement levels remain unsatisfactory, gender differences in participation are resisting change and the cost of provision remains so high that many pupils throughout the world are denied access to effective science teaching. Scientific literacy remains a long-term goal despite the widespread inclusion of science in most national curricula. The output of qualified secondary science graduates, who can become the technicians, engineers and researchers of the next generation, is often inadequate.

If much has been written on the problems of teaching science and about the nature of more relevant science curricula, little has been said about the problems that confront national planners and policy-makers. Decisions have to be made about the use of public resources: how much should be invested in science for all and how much in science for future scientists, for example. Value for money is essential. Curriculum changes have been profound and frequent and knowledge on what is really happening at the school and classroom level is scarce.

This book addresses the question of how best to plan investment in science at secondary level. It is based on a five-year research programme which was conducted at the International Institute for Educational Planning during its Fifth Medium-Term Plan. This programme included studies in 12 countries, several state-of-the-art monographs written by acknowledged experts, and detailed country-case studies. This output has been discussed in national and regional forums, leading to fruitful discussions with national policy-makers. The result is an extensive review and analysis of the main issues confronting the planning of science on the eve of the 21st century. It aims at helping policy-makers and planners to formulate policies with positive far-reaching impact and to provide them with the tools to review outdated practices and monitor implementation of their policies.

This report has a threefold interest: it provides the reader with original insights into the way science education is organized in different countries, delivered at school level, including the school leavers' prospects in further education and employment; second, it examines the main issues that can be identified in planning and implementing science education policy – specialization, selection, unequal access and gender balance, the role of practical activity, assessment issues, language policy, and teacher training provision – and deliberates on cost-effective ways of delivering high quality science education; finally, it emphasizes the need to improve the information base on science education to allow for its regular monitoring. It will, therefore, interest a wide audience including educational planners in developing countries, decision-makers and policy analysts working on human resource development in a wide range of countries, donor-agency programme officers, academics in science education, staff of science teacher training colleges, and all others interested in education, development studies and science policy.

I would like to take this opportunity to express my deep thanks to the UNDP for the financial support given to this project.

Jacques Hallak
Assistant Director-General, UNESCO
Director, IIEP

Acknowledgements

This report is the synthesis of a research project which was comprised of a number of national monographs, detailed country case studies and state-of-the-art reviews. Their authors are too numerous to be listed here, but we would like to express our gratitude to them all. We should like to thank specifically the leaders of the teams with whom we conducted the detailed case studies: Sharifah Maimunah in Malaysia and Messrs Radi and Hddigui in Morocco. Without their hard work and their dedication, this synthesis could never have been written.

A special thanks goes to the education authorities of Malaysia and Morocco for the interest they took in the case studies in their respective countries and for the material, human and financial support they provided for their implementation.

Our thanks are also due to Jacques Hallak, Assistant Director-General, UNESCO, and Director of IIEP, and T. Neville Postlethwaite, Professor (Emeritus), University of Hamburg, as well as to all those who made comments on the first drafts of this synthesis.

Finally, our gratitude goes to Anne Pawle who read, reread and edited the various versions of this manuscript.

FC
Paris, February 1997

Contents

Introduction

An explosive growth of information and scientific knowledge, resulting in new products and processes, has characterized the twentieth century. It has brought about the transformation of the way of life of billions of human beings. Not all societies, however, have received the global benefits of this scientific era, and one of the major challenges facing human resources planning today is in how to deal with the uneven level of scientific and technological development in the various countries. The ability to select, apply, maintain and ultimately develop appropriate science and technology products is crucial to the process of modernization. An element which curtails progress in most developing countries is a serious shortage of science and technology specialists.

This book addresses the question of how best to plan science investment at secondary level and discusses the issues that planners of science education in developing countries will face when trying to meet the challenges of the twenty-first century.

Educating high-level science specialists alone will not enable modernization to take place nor poverty to be reduced. Overall scientific literacy and the spreading of a mature and appropriate scientific and technological capacity has to be achieved if development is to be sustained. As many children and young people as possible should be educated in science, not only to prepare them to become better workers, but primarily to help them understand their environment and enable them to act upon it. Young people everywhere have to be prepared to face a future that will be substantially different from the present. As citizens they have to be able to participate in an informed way in some of the technological and social debates of the future.

As early as the 1960s, many developing countries embarked on programmes to support the development of science education at secondary and higher education levels. Since then, investment in science education appears in almost every national development plan as a priority area. Curricula have been revised; many teachers have been trained and much time and money have been invested in pre-service and in-service programmes; considerable resources have also been invested in laboratories and in equipment to support science teaching based on practice.

More than 30 years later, the results have not always met expectations: patterns of achievement, as illustrated in examination results and in international surveys, are often worrying, particularly amongst average and below average students. Scientific literacy throughout the population remains a distant goal in most countries, despite widespread inclusion of science in national curricula. Growing anti-science sentiments, which have started to develop in some societies and political circles, show that large segments of the population are uncomfortable with science and perhaps with modernization in general. In many countries, students are reluctant to study and, even more, to specialize in science, a subject which they consider difficult and abstract. Girls are particularly under-represented in such studies, and this may have lasting consequences for development. The inadequacy of the output of qualified secondary science graduates often leads to poor success rates in science studies at higher education level. Finally, science graduates, who used to be protected against unemployment, are beginning to face serious recruitment problems in a number of countries. Some of them, unable to find a job in the public sector, remain unemployed for a long period of time, and this, in a way, demonstrates that the expected results have not been attained. This is especially of concern where such unemployment coexists with the continued employment of expatriates in science-based employment.

There are many reasons for these unsatisfactory results, some of them lying inside the educational sphere, and others outside. Much has been written on science education, on the curricula, the learning processes, the teaching methods, and best practices and innovations in assessment methods in science. Little has been said of the problems and dilemmas that policy-makers and planners face when making their choices and allocating scarce resources: what should be the priorities – science for all or the training of future scientists? To which schools should the most effective teachers be allocated? Who should benefit from the best facilities? Where should investment be focused: on lower or upper secondary pupils? What are the costs and advantages of different curricula options, and of introducing technology?

Decision-makers across the world are concerned with ways and means of improving levels of achievement and enhancing the outcomes of science education. They are also concerned with how to make better use of their scarce resources and how to obtain value for money.

The first challenge is to evaluate the state of science education in different countries. Information is often lacking. In many developing countries existing statistics do not provide sufficient information on the number of pupils who study different types of science at secondary level, how many are likely to continue studying science at higher level, and whether the output of the school system is in balance with the requirements of higher education or the labour market. Information on teaching conditions, and on the quality and the cost of providing science education at different levels, is even more scarce.

One of the objectives of the International Institute for Educational Planning's research project, on which this book is based, was to establish the conditions of science education in a range of developing countries. A number of studies were undertaken and monographs written which specifically aimed at:

(i)   establishing the conditions of science education at secondary level in countries at various levels of economic development;
(ii) assessing the factors which may have prevented the investment made from having the expected outcome;
(iii) identifying the most promising strategies for providing science education in a more effective manner;
(iv) measuring the impact of science education on human resource development;
(v)   developing techniques and indicators of use to the planner in assessing science education provision.

Studies and discussions undertaken in the framework of this research programme brought to light the fact that in many countries there is a dearth of communication and co-ordination between the different administrative units concerned with the organization and delivery of science education: curriculum developers too often develop their curriculum without taking into consideration the country's resource constraints or the teaching conditions prevailing outside metropolitan centres. Universities and teacher training colleges may do very much the same, sometimes ignoring recent curricula reforms. Examination Boards continue to put the emphasis in their examinations on recall questions. This book will try to bridge the gap between the different actors involved in the organization of science education.

The research focused on secondary education, since it is at that level that most initiatives have been taken and where human resources needs have been identified as most critical for future development. Primary science is certainly an important area of concern, but it is at secondary

3

level that the majority of pupils experience formal science education, usually for the first and last time before entering the labour market. Providing good science education at that level is considered crucial for a variety of reasons. Firstly, it consolidates and strengthens what pupils may have learnt at primary education level. Second, it provides a sound basis in science for those who will educate the next generation of primary school students: a majority of primary school teachers receive no other training in science than that which they experienced at secondary level – generally speaking, in a non-science stream. Third, good secondary science is essential for the preparation of middle-level scientifically and technically based workers who will either go straight into employment or follow technical and vocational courses later on. Finally, it is at the secondary level that the selection and training takes place of the future scientific and technical élite who will proceed to university level and professional jobs. Obviously, the better the science education they receive at secondary level, the easier it will be for them to learn at higher levels.

The work undertaken included:

- Two state-of-the-art reports on the qualitative aspects of the teaching of science, one based on Anglophone and the other on Francophone literature.
- An international survey on the conditions of science education provision in academic secondary education – this survey, which covered 12 developing countries, collected information on the organization of science teaching, the number and profile of the pupils studying different science programmes, the methods of selection, guiding and assessing pupils in science, the curriculum organization and the teaching/learning conditions, and on certain elements of cost in each country.
- Two in-depth case studies on two middle-income countries: Malaysia and Morocco. These two countries were selected because they present an interesting and contrasting picture of the problems that countries face in the provision of high quality science education. The two countries have a sustained commitment in common to provide high quality science education and a good level of resource provision. In spite of this, the quality of their science education and the level of pupils' achievement are not wholly in line with what may be expected. Beyond these common characteristics they differ on several grounds. Malaysia's organization of curriculum is influenced by the English model, while Morocco has features of the Francophone model. Malaysia is one of the fast-growing

economies of South-East Asia characterized by an acute shortage of scientifically trained personnel. It also has to deal with a decline in the number of pupils studying science. Morocco, on the other hand, has made considerable investments in science education in order to localize its high-level workforce, but now faces a situation of over-supply of science graduates. Whatever the problems these countries are facing in their science education provision, it cannot simply be attributed to a lack of resources. Lessons can therefore be derived from their experience which may be of interest to other countries.

The case studies included five elements – a baseline study, a review of the destinations of secondary science graduates, a survey of a sample of science teachers and school heads, case studies of selected schools, and an analysis of assessment data. The baseline study drew together secondary data on the flow of students, enrolment rates, provision of teachers, curricula patterns, and resources available. The study of destinations of secondary school graduates involved the accumulation of statistics on the intake to post-secondary institutions and, in the case of Morocco, an analysis of the enrolment file of students throughout their higher studies. The case-study work undertaken in Malaysia involved the collection of very detailed information on the organization and delivery of science education. The methodology followed is presented later in the report.

• A series of monographs investigating a number of key issues in the delivery of science education, such as the use of science kits, girls in science education, environmental education, efficiency and the desirability of special science secondary schools, the teaching of technology and, last but not least, the training of science teachers.

The main findings of the IIEP's research programme have been organized into six chapters.

What is the contribution of science education to development? This is the question which is investigated in *Chapter I*. Although there is a close association between the number of people trained in science and technology and simple indicators of economic development, there are some patent counter-examples.

Chapter I outlines some of the conditions necessary for the development of a scientific and technological capacity. It also discusses an issue which has appeared in recent literature on human resource development: can science education be a substitute for technical and vocational education?

Chapter II summarizes the findings concerning the organization and condition of science education in the various countries covered by the project. Information is based on the IIEP survey and on its case-study work. It looks at the organization and coverage of science education in different countries, the curriculum patterns, the teaching conditions, the levels of achievement, some elements of costing, and the destination of school leavers in further education and in employment.

Chapter III reviews, one by one, the most crucial issues involved in the planning of science education provision. These are concerned with: flow control – how many students should be specialized and how early? How should students in science be selected? How can science be made more attractive to an additional number of students – all students, and girls in particular? What resources are needed to support the present curriculum trend which sets out to attack the world's problems through environmental and technological education? Is practical activity absolutely essential for proper science teaching? How could teachers be better trained?

Chapter IV reviews some of the strategies which are currently used to provide science education in a cost-effective way. It looks in detail at the advantages and disadvantages of sustaining special science schools; it also explores the cost implications of laboratory provision and science kits, the improvements that can be expected from the development of appropriate science learning materials and their delivery, and from effective support mechanisms for teachers at school level.

The research experience shows that planners and policy-makers have very little data on which to base their decisions.

Chapter V draws some methodological lessons from the research work and suggests ways and means of improving the information base for the monitoring of the status of science education provision, and for identifying interventions which are needed.

Chapter VI summarizes the main findings and draws up some tentative policy recommendations which will be of use to planners and decision-makers in developing countries.

Chapter I
Science education and development

Science and technology have transformed the world, mostly for the better. Various applications of science and technology have led to dramatic decreases in infant mortality, the doubling of life expectancy in many countries, the decrease of epidemics and, to a great extent, the disappearance of large-scale famines which can decimate populations. Throughout the world more and more people are able to participate in mass consumption and benefit from products not previously available; new sources of energy have been developed that sustain cities and higher levels of productivity. Many of these developmentally desirable outcomes would be difficult to achieve without economic development and there again science and technology appears indispensable. Technical change is considered essential for increasing productivity, for competing in international trade and for attaining sustainable economic growth.

It is true that development has brought with it a number of problems such as pollution, deforestation, land degradation, or new forms of poverty in cities. It would be easy to illustrate however that science and technology could often provide solutions to these problems should a real will exist to address them. Rather than a lack of faith in the potential of science to overcome poverty, disease or even environmental degradation, the concern is that many developing countries continue to have a limited capacity to adapt the technology they buy to their needs and to widely assimilate useful technical knowledge, know-how and related practices. A related concern is that science and technology contribute to increasing disparities between developed and developing countries, as the latter may not be in a position to invest as heavily as is desirable in new and more advanced technology.

A case in point is the present 'information revolution' that is taking place. It holds out the prospect of empowering peoples throughout the world through greater access to knowledge and vastly improved means of communication. At the same time, the new information and communication technologies, which affect all sectors of the economy, are likely to

create serious and long-lasting disparities between those countries which
use them and are able to produce goods and services related to information and those which do not. The latter run the risk of being marginalized
for a long time.

The revolution of information technology and the consequent
globalization of the economy are however a reality. The best chance of
grasping the benefits of new developments and minimizing the losses
seem to be in investing in domestic capacity to capture and exploit
opportunities towards desired developmental ends. This implies systematic investment in the human resource base which has the scientific and
technological competency to apply existing and new technologies to
satisfy needs.

Conscious of the need to catch up, many developing countries have
indeed invested heavily to create their own scientific and technological
capacities. These efforts have included, among other things, the training
of large numbers of scientists and engineers, both locally and abroad. This
chapter reviews some empirical evidence on the number of scientists and
engineers that have been trained in different countries in the past 25 years
and relates this to some statistics on the evolution of GNP and GNP *per
capita*[1] during the same period. This leads to the conclusion that investing
in science is a necessary but insufficient condition for development to
take place. This chapter then discusses some of the factors which may
explain why the investment in human capital did not necessarily have the
impact it was supposed to have on economic growth, and reviews briefly
the debate on models of scientific and technological development.
Drawing among other things on the recent experience of high-performing
Asian economies, some of the conditions associated with successful
growth strategies are outlined. Finally, there is a discussion about the
priority to be attached to secondary science education versus vocational
education. This is seen as important, because science education is often
presented as being a potential alternative to school-based vocational
education, when it comes to preparing students for entering the world of
work.

1. The well-known limitations of using GNP and GNP *per capita* as a proxy
    for economic development are recognized. In the absence of a widely
    understood alternative we continue to use it for this summary discussion.
    The human resource development index is not appropriate since the
    primary concern is with economic development.

8

1. Facts and figures: enrolments in scientific education and economic growth

Statistics on the evolution of enrolments in higher education by field of study exist from 1970 to 1990. The number of students enrolled in science and engineering studies per 10,000 inhabitants, an indicator which measures the annual effort made to train high-level specialists in science,[2] has been computed. This can be related to the growth of GNP *per capita* during the same period. Data were available for some 110 countries.

Table 1.1 summarizes the value of the indicators in 1990 and in 1970 in selected countries. In 1990, the countries which trained the largest numbers of science specialists[3] per 10,000 inhabitants were: Korea (114), Finland (116) and Germany (104). These numbers were greater than for the United States (91), Japan (44) and the majority of European countries: e.g. United Kingdom (60), France and Italy (65). Various developing countries seemed to be training a proportion of young people in science which is comparable with those in some developed countries: e.g. Algeria (41), Hong Kong (50), Jordan (41), Philippines (56), Syria (48), Mexico (48), Panama (44), Chile (66), Colombia (42), Peru (51), and Uruguay (40). This would appear to be justified by the need to increase the stock of scientists and technologists to levels closer to those of developed countries.

In 1970, on the other hand, the countries that trained the largest proportions of youngsters in science were the Eastern European countries, e.g. USSR (85), Czechoslovakia (37), Bulgaria (55), Poland (49), followed by the Netherlands (51), Israel (40), and Japan (36). These countries were well ahead of the USA and other European countries, most of which already had a substantial number of science-trained individuals in the labour market. At that time, several developing countries were already training relatively large proportions of their youngsters in science: e.g. Chile (23), Argentina (20), Singapore (29), Hong Kong (25) and Korea (20). A difference between Asian and Latin American countries is however worth noting. Asian countries attained such large proportions primarily as a result of the high proportion of their student body specializing in science, while Latin-American countries attained similarly high proportions because of their relatively high enrolment ratio at higher education level in general.

2. A proxy for the enrolment ratio.

3. Students enrolled in engineering, mathematics, natural science, computer science at higher education level. Medical students have been excluded.

Table 1.1. Indicators of scientific density in selected countries – 1970-1990.

Regions	Countries	GNP per capita 1990 (US$)	Science students per 10,000 inhab* 1990	GNP per capita 1970 (US$)	Science students per 10,000 inhab * 1970
Africa	Botswana	2 040	5.84	320	0.74
	Cameroon	940	7.79	246	0.63
	Congo	1 010	5.35	441	1.00
	Egypt	600	12.32	250	12.77
	Ghana	390	1.62	580	0.96
	Kenya	370	2.28	206	0.69
	Mauritius	2 250	3.69	550	4.15
	Morocco	950	28.86	477	0.62
	Nigeria	270	4.48	309	0.62
	Senegal	710	5.54	350	1.47
	Togo	410	2.55	204	0.63
	Tunisia	1 420	17.95	660	4.64
	Zambia	420	4.00	398	1.27
Asia	Bangladesh	200	7.73	100	3.42
	Hong Kong	11 540	50.19	1446	25.84
	India	n.a.	n.a.	139	3.08
	Indonesia	560	14.83	196	3.84
	Korea	5 400	114.27	308	20.74
	Malaysia	2 340	16.81	684	4.27
	Philippines	730	56.80	356	16.64
	Sri Lanka	470	9.33	178	1.49
	Thailand	1 420	24.28	292	2.07

Table 1.1. contd.

Latin America	Argentina	n.a.	n.a.	1 313	20.24
	Bolivia	620	40.72	340	6.97
	Brazil	2 680	19.46	860	9.68
	Chile	1 940	66.49	995	23.19
	Colombia	1 240	42.12	534	9.63
	Honduras	590	23.96	357	2.04
	Mexico	2 490	48.75	913	14.90
	Uruguay	2 560	40.29	1 341	2.65
OECD Countries	Canada	20 450	67.64	6 109	37.13
	France	19 480	65.03	5 117	23.31
	Germany	20 750	104.03	6 039	30.37
	Japan	25 430	44.54	3 111	36.44
	United Kingdom	16 070	60.12	3 528	35.26
	United States of America	21 700	91.31	6 884	8.33
Eastern European Countries	Bulgaria	2 210	81.37	1 764	55.02
	Former Czechoslovakia	3 140	49.95	3 292	37.68
	Hungary	2 780	23.19	1 909	30.62
	Former USSR	6 000	74.60	2 207	85.48

*Number of students per 10,000 inhabitants studying science and engineering at the tertiary education level.
n.a.= non available.

Source: Computed from UN Statistical Yearbooks and UNESCO Statistical Yearbooks.

African countries had relatively fewer students in science and technology-related studies both in 1970 and in 1990. Characteristically this was a result of lower general enrolment rates and smaller proportions specializing in these fields. The unit cost of these science and technology students was typically a greater proportion of GNP in Africa than in Asia. This did not necessarily imply a lower level of commitment in Africa; it partly reflected severer constraints on investment in human resources.

To check whether there was a relationship between the number of students enrolled in science and technology education (per 10,000 inhabitants) and the GNP *per capita*, a number of regression analyses were undertaken. Worldwide, a fairly strong relationship was found, at least in 1990, between the number of students studying science per 10,000 inhabitants in a country and its GNP *per capita* (see *Figure 1*). Generally speaking, the richest countries can afford to enrol a larger proportion of their population in higher education, and in science-related studies, than can the others. Doing the analysis separately by groups of countries it appeared that the regression coefficient was higher for the poorer countries than for the richer ones (those with a GNP *per capita* above $5,000). In the richer group, Eastern European countries were an anomaly in the general pattern since they had a much higher ratio of students studying science than Western European countries and yet a lower GNP *per capita*.

A significant regression coefficient between the indicators for the same year does not of course indicate that enrolling a large number of students in science will lead to a higher GNP *per capita*. If there is an effect it will be subject to a time lag. It is the people who were trained 15 or 20 years ago who could possibly be responsible for some part of the present economic growth. In order to test the influence of investing in higher science studies on economic growth over time, another set of regression analyses was undertaken, linking the growth rates of the GNP on one hand and the GNP *per capita* in 1990 on the other to the proportion of the population studying science in 1970.

The results were equivocal. There seems to be no obvious relation between these indicators. To take the case of the five developing countries mentioned above as having a fairly large number of students studying science, relative to their population in 1970, three (Korea, Hong Kong, Singapore) had witnessed an extremely rapid economic growth – Korea, for example, saw its GNP *per capita* more than double from 1970 to 1990 – while the two Latin-American countries had experienced very moderate growth during the same period.

Figure 1.1 GNP per capita and the number of students in science and engineering at higher education level (1990)

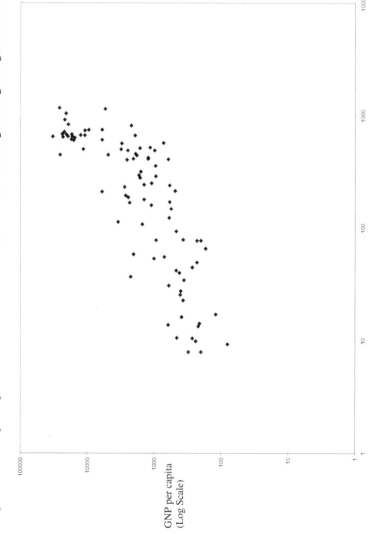

GNP per capita
(Log Scale)

Number of science and engineering students at higher education level per 100 000 inh. (Log Scale)

The case of the Eastern European countries' economies, which had not grown rapidly despite heavy investment in science and technology education, is a reminder that macro-economic events and political instability may negate whatever benefits stem from investment in human resources. This is one possible reason why a relationship did not clearly emerge. Another possibility may be that the time period selected was too short; unfortunately, longer time-series data were not available for the majority of countries.

If investing in science education is a necessary but not a sufficient condition for economic growth, what are the other conditions that may need to be met?

2. Debate on the models of scientific and technological development

There is abundant literature dealing with the problem of creating scientific and technological capacity in developing countries and with the link between the existence of scientific resources and development (IIEP, 1984; Katz, 1984; Sagasti, 1989; OECD, 1992; Salomon et al., 1994; Wad, 1994). This is not the place for an extensive review. The experience of Latin-American countries, however, is informative, and will be taken as a case in point. At the end of the Second World War, most had at their disposal a viable scientific community as well as an emerging industrial base. Despite this, many were unable to use this capacity to foster sustainable economic development.

Part of this failure was first attributed to the orientation of the scientific community, many of whom favoured focusing their work on research of interest to comparable scientific communities in advanced countries. In the 1960s, however, many of the governments became more conscious of the role that science and technology could play in socio-economic development and in national independence. As a result, a series of national institutions intended to formulate national scientific and technological policies to finance research and development programmes were created. A large number of state enterprises meant to implement these policies in key sectors of the economy were also established and trade barriers were introduced to protect production in their infant industries. In what became the 'import substitution' development model, the state had a large role to play both as a promoter of the technological policy (through incentives) and as a direct actor in its implementation (through state firms).

These policies led to some definite achievements: production developed very rapidly; a real physical infrastructure was created and the industrial sector was reinforced; social services were also introduced and expanded. These achievements, however, were often restricted to some specific industries, to specific regions, and to certain groups; the benefits did not trickle down to the rest of the economy. Enclaves of highly developed industries in an economy and society which generally remained poor became a common feature of the industrialization process of countries like Brazil, Venezuela, and Mexico. Similar experiences have also been observed in other developing countries such as India and Iran. In Latin America, the severe depression in the 1980s, the decline in GNP *per capita* and growing poverty demonstrated the limits of such achievements. They led to the redefinition of economic and technological policies.

In general, governments overestimated their capacities, and that of their institutions, to plan and implement a policy leading to growth based on investment in science and technology. It also seems to have been the case that governments underestimated the significance of cultural factors in mobilizing scientific and technological resources, and the difficulties that arise in adapting transferred technology when the domestic base of scientifically competent human resources is limited. It may very well be that Latin-American governments, and some other LDCs in Asia and Africa, paid too much attention to attempts to create leading-edge industries and not enough to the reduction of disparities in income and access to education between rural and urban areas and between social groups, thus exacerbating development problems related to distribution. In education, levels of investment at the tertiary level in general were sometimes disproportionate to the benefits they delivered. In a desire to catch up fast, too much attention was paid to the training of top-level specialists, while enrolment at primary and secondary levels remained low, and early drop-out levels high. Primary and secondary science hence remained underdeveloped and this may have slowed down the process of learning and gradual assimilation of science and technology by the overall population.

It takes more than one generation for scientific knowledge and technical know-how to be assimilated widely throughout a society. Without a sustained emphasis on such learning, gradual assimilation by substantial proportions of the population, and maturation of mechanisms to apply knowledge and skills, superficial transfers of technology may occur which are more transplanted than genuinely transferred (Salomon, 1984). They may have only a marginal impact on society and development, and fall short of providing the conditions under which sustained growth may occur.

15

In the 1970s and 1980s, models of development that stressed technology transfer were set in contrast to those which rejected imitative development in favour of domestically defined investment priorities related to appropriate and indigenous science and technology. Alternative technology approaches favoured greater focus on rural development, small-scale innovation, and low-technology solutions to address the development needs of the poorest parts of society directly. The debate was extensive and absorbed much space in the literature (cf. Herrera, 1978; Arghiri, 1980; Fransman and King, 1984; Katz, 1984; Sagasti 1989; OECD, 1992; Salomon and Lebeau, 1993; Wad, 1994). Several factors have changed, however, since the 1980s which have altered the parameters of this debate.

In brief:

- Market-based development strategies have widely displaced centrally planned, protectionist and self-reliant models of development and many countries, which previously embraced the latter, have adopted the former.
- It has become increasingly clear that capital is not a constraint to development in most countries; it comes from various sources and is available where comparative advantages exist.
- The globalization of the world economy has been rapid and unavoidable; it has become far easier to locate production anywhere in the world where natural resource endowments and appropriate human resources exist.
- Locating to the lowest labour costs, especially in export manufacturing, has diminished in importance in relation to other factors such as political stability, quality of the labour force and its educational level.
- Industries with a high value-added component have become very knowledge intensive and most depend on an adequate science and technology base.
- Information flows freely and rapidly internationally through the new communication technologies creating many new possibilities for growth in production and employment in developing countries.

Although the diffusion of appropriate technologies undoubtedly remains important to rural development and the improvement of living conditions amongst marginalized groups, few now argue that it is possible to remain outside the mainstream of economic globalization. Viable strategies to generate an industrial base and create the basis for mass consumption of goods and services depend on success in producing tradeable commodities. Changing the balance from the export of primary products (characteristic of low-income economies) to manufactured goods (associated with countries that have seen substantial increases in real wages and living standards) depends on the adoption of best practices by increasingly skilled workforces. This implies the understanding of and access to global science and technology and its applications. It is being increasingly argued that "for the late starter, the development process has to be a learning process rather than a discovering or an inventing process" (Dore, 1989:102).

It now seems widely acknowledged that no country, except possibly the largest, can realistically attempt to develop a comprehensive indigenous science and technology base in every possible field, just as industrial investment is better targeted on areas where global competitiveness is a realistic prospect rather than a wishful dream. This often implies that it is necessary to develop the capacity to buy, select, maintain and use new technologies and to be able to produce in areas of activity which reflect local resources and comparative advantages. Small countries in particular need to identify affordable technologies and areas of activity where they can have such a comparative advantage. Countries like Korea, China/Taiwan, China, Hong Kong and, more recently, Malaysia, Thailand, Indonesia, or Mauritius have achieved this in differing measures and have succeeded in developing very quickly. Growth in GNP *per capita* of some of these countries has been so rapid that the gap in income and human development indices has been closed, to an unprecedented extent, between themselves and established industrialized countries.

3. Lessons from the experience of Asian high-performing economies

A number of factors emerge from the recent experience of high-performing Asian economies in contrast with those of less economically successful countries. They are worth listing as factors which possibly contribute to success.

Sound macro-economic policies

Most analysts agree that these countries have pursued a set of macro-economic policies which have underpinned growth (see, for example, World Bank, 1993). These have several features:

Exports of manufactured goods have been favoured, especially in those sectors where value added is high (which are often also sectors which are science- and technology-based); production and export of non-processed primary products have fallen, at least in relative terms.

Industrialization started with light industry and progressively moved towards more and more knowledge-intensive products, and 'intelligent' production methods have been adopted which reorganize means of production which value competency and distribute responsibilities throughout the labour force.

Many of the most successful developing countries have high saving ratios. Contrary to Latin-American and African countries, they did not borrow heavily to finance their investments. Where they did, safeguards existed to guarantee adequate returns. Much development was therefore supported by domestic resources. Political stability was also a key factor in attracting both private and public foreign investment where this took place.

Reliance on several sources of technology acquisition

All the East Asian newly industrialized countries have followed diverse technology acquisition and development strategies which have been accompanied by substantial investment in education and training in general, and science and technology in particular. Technology has been acquired and refined through creative adaptation of existing mature technologies, joint ventures and/or purchase of licences and patents, and in some cases through direct foreign investment. A large number of students have studied abroad and returned, especially in science and engineering[4]. Concurrently, they have invested heavily in research and development. Enterprises themselves have been responsible for a great deal of the R&D conducted[5]. Governments have adopted broad strategies that have encouraged enterprises to assimilate and adapt new technologies.

4. In 1990, 35 per cent of all undergraduate and graduate Korean students studying in the USA were enrolled in engineering and another 35 per cent were enrolled in natural sciences. Nearly 50 per cent of all doctoral degrees in these fields which were awarded to Korean students were awarded in the USA. Similarly, 45 per cent of all undergraduate and graduate Chinese students from China/Taiwan studying in the USA were enrolled in engineering and another 51 per cent were enrolled in natural sciences. Seventy-three to 81 per cent of all doctoral degrees awarded in the same fields to Chinese students from China/Taiwan were also awarded in the USA (National Science Foundation, 1993).

5. In 1987, in Korea, nearly half of all researchers were working in a private enterprise.

Balance between government and market regulation

The design of the policy and the orchestration of the overall strategy was achieved by a balanced mixture of central government intervention through establishing parastatal enterprises, providing adequate incentives, promoting selected sectors for investment and supporting innovation in these areas, and market competition to increase competitiveness (Freeman, 1989).

Heavy investment in human capital

It is striking that these economies have chosen to invest heavily in the development of human capital progressively in response to – or in anticipation of – the changing needs of their economic and industrial development. Most of the high-performing countries in Asia started investing first in mass primary education; only after primary schooling was close to being universal did the growth of enrolments in their lower secondary schools increase, to be followed by expansion at upper secondary school and university levels (Caillods, 1994).

The case of Korea is illustrative (Jong Ha Han, 1993). Korea expanded its primary education in the 1960s, when the country's development was based on labour-intensive light industry with a limited need of skilled manpower. By the end of the 1960s, the country had achieved universal primary education and was starting to expand lower secondary education. By the 1970s, the first stage of secondary education was open to everybody. At that time, the national development strategy began to favour the development of heavy and chemical industries, and more emphasis began to be placed on technical and vocational education. Large numbers of technical and vocational schools were opened in the 1980s, along with vocational institutes offering shorter and more vocationally oriented courses. The emphasis was not only on school education – all substantial enterprises were encouraged, through tax incentives, to set up their own training centres. All enterprises of more than 300 employees were required to train a proportion of their workforce. In the 1990s, with industry shifting to more sophisticated products and microelectronics, the emphasis became placed on higher technical courses and advanced studies. Efforts have also been made to popularize science and technology throughout the population. A nationwide movement was created to encourage the application of scientific principles to most aspects of national life, with the co-operation of the academic and industrial communities and the mass media.

Much of the above would not have been possible without the cultural characteristics of the East Asian societies: the value attributed to education, the conformity to community standards, the comparatively weak notion of 'self' in hierarchical configurations (Hicks and Redding, 1983 and Cheng Kai-ming, 1991), etc. The experience of the high-performing Asian economies may not be easy to replicate in countries which start without similar cultural dispositions. The economic context has changed as well. Latecomers like China, some South-East Asian countries now and others in future, may very well have to vary their development strategies to reflect the existence of these newly industrialized economies and modify some of the approaches that proved successful for them. The basic principles mentioned above are, however, worthy of considering seriously by countries wanting to build up a scientific and technological capacity.

4. Science or vocational and technical education?

The experience of high-performing economies suggests that the creation of a scientific capacity requires systematic investment in human resource development, giving some priority to basic education to attain widespread scientific literacy, while investing selectively at upper secondary- and higher-education levels, particularly in science and technical streams. What priority should be given at secondary level to general academic or to vocational and technical education is then an important issue. What would be a most cost-efficient strategy for countries with limited resources: providing a basic science and technology education in all secondary schools or maintaining special tracks for those who will enter the labour market?

Some studies based on historical data explored this issue (Hage, Garnier and Fuller, 1988; Garnier and Hage 1994; Walters and Rubinson, 1983). They tried to measure the impact of different types of provision on the economic growth of France, England and Germany after the Second World War. While the general contribution of education to economic growth has been extensively researched since the well-known studies by Denison (1962) and Schultz (1961), Garnier and Hage (1994) have recently tried to include in their analyses judgements of the quality and of the content provided. These suggest that the extent of teaching of mathematics, science and technology, the number of hours spent on these subjects and the quality of the teaching process may have contributed substantially to the economic growth of England and France. They also argued that technical education played a major role in contributing to economic growth in the case of France and Germany (secondary technical education in the case of France and higher technical studies in the case of Germany) but less so in England (Garnier and Hage, 1994). In a similar vein, a number of American economists attributed the 'relative decline' of the USA in some industries and in R&D investment to the condition of the public education system in the USA and to the comparative lack of basic skills, particularly in the technical area, acquired by the general population (Harvey Brooks; Kearns and Doyle, 1988)[6]. In Europe, the superior economic performance of Germany is often attributed to the country's strong vocational and training system. Prais and Wagner (1986) argued that while this was the case, it should also be attributed to the fact that there was within the German secondary education system a much stronger work-orientation programme (than in Britain), more concern for raising the achievement of the average pupils in maths and science, even in the pre-vocational and in the vocational streams (resulting in a higher mathematical competence in Germany), and generally stronger links between secondary schools and subsequent vocational schools. In other words, there was a complementarity between mathematics and science studies and the teaching of vocational subjects, rather than an opposition or separation between them.

Most continental European countries retain a school-based vocational and technical education system in parallel to general academic streams. What remains the object of a debate in these countries is: the content of the most appropriate courses, and the delivery methods (emphasizing alternance between schools and firms) (OECD-CEREQ, 1994; Greffe, 1997). Many of the newly industrialized countries also maintain sizeable vocational and technological streams in parallel with their academic streams. In a number of Latin-American countries the general quality of the education provided and the quality of science education in particular, was often better in technical schools than in the general school system[7] (Leite and Caillods, 1987; Gallart, 1990; Calzadilla and Bruni Celli, 1994). The same could be said of some Eastern European countries and Russia (De Moura Castro et al., 1997). As a result, vocational and technical school graduates there can find a job more easily than academic

6. Bishop (1989) attributed the slow-down in the USA productivity growth during the 1970s to the test score decline between 1967 and 1980 – where test scores were taken to measure general intellectual achievement, i.e. all competency which contributes to productivity in most jobs.

7. Argentina and Brazil are now considering, however, pushing such technical and technological studies to the post-secondary level.

school graduates; they earn higher salaries and many of them have a good chance of entering higher education.

Inevitably, however, experience varies from country to country and with the level of economic development, the size of the industrial base, and the quality of the education provided in different streams. In the 1960s and 1970s, conventional wisdom encouraged many developing countries to invest in vocational schools in order to give the workforce the various skills thought to be required by economic development, and to improve workers' adaptability and productivity. These initiatives have not been conspicuously successful in the low-income countries suffering from serious resource constraints; vocational schools turned out to be costly, often offering programmes of questionable quality and producing graduates at a high cost who enjoy no special advantages in seeking employment. Such schools were frequently thought to be unresponsive to the changing requirements of the labour market. The experience has led many economists to argue for reductions in the scale of government-funded, institutionalized vocational education in favour of on-the-job training and programmes more closely related to the workplace and sponsored directly or indirectly by employers who can identify their needs (Middleton, Ziderman and Adams, 1993). The suggestion has also been made to replace separate school level vocational programmes by more intensive, more applied and technology-oriented science courses within general education.

Increasingly, the tendency is to delay entrance into vocational education and training programmes until the end of lower secondary after eight to nine years of basic education. This is partly in recognition of the fact that it is easier to teach vocational skills to students who have a good foundation in mathematics, science and technology, as well as in languages. After nine years, school leavers should have the general skills essential for performing many tasks as well as the general foundation upon which job-specific knowledge can be built. It would be dangerous to conclude however that any trade can be learnt on the job. For several trades, attendance at organized training sessions remains essential to learn the practical but also the technological and technical background (De Moura Castro, 1995).

The implication of this discussion and of the examples above is that the choice between science and vocational education is a false one: if training is easier and faster if students have had a good basic education in science, then this should be a prerequisite for entry to higher level training in related fields. Specific job-related skills still have to be taught, depending on the occupation in specific streams, courses, or on the job. The choice will depend on the structure of the economy, and historic and culturally determined patterns of differentiation within the school system. *Whichever the case, the teaching of mathematics and science within training courses and programmes should be reinforced.*

Conclusion

Educating large numbers of people in science (and technology) may be a necessary condition for a country to acquire a competitive technologically based industrial capacity, but it is not sufficient. Other conditions need to be satisfied, which probably include a stable political environment, competent public administration, scientifically aware management capability, and an appropriate macro-economic policy environment.

The creation of a scientific capacity requires a dual-pronged educational investment strategy which balances needs to extend scientific literacy in general with the supply of science and technology specialists at high levels. Either one alone may be insufficient. The importance of investment in basic education is clear and has been a central feature of the rapidly developing economies. Basic education includes the assimilation of fundamental science thinking skills and knowledge. High-level scientific capacity is needed to ensure that enterprises should be able to select, adapt, maintain and incorporate the technology they acquire. This is dependent on the existence of specialized scientists and technologists able to work in productive industry as well as on research.

Training of the workforce through formal training programmes, on-the-job training and experience is desirable. This is likely to be easier the better the basic education and scientific foundation established at the school level. Hence, science education at secondary level should be regarded as a prerequisite for entering vocational and technical courses or further education and training programmes, and a complement to be reinforced in these programmes, rather than as a substitute.

Chapter II

The state of science education in different countries

The aim of this chapter is to review the condition of science education in a number of developing countries. Unfortunately, information on this subject is sparse. There are few studies available, most notably the Second International Science Study (SISS) of the International Association for the Evaluation of Education (IEA, 1991, 1992). This study included a very limited number of developing countries and used data from 1983/1984. Since then, economic recession has been widespread and has had an adverse impact on educational investment in numerous Arab, African and Latin-American countries. Adjustment programmes have resulted in stagnation in real terms of the resources available for education. Science education, especially since its costs are often relatively high compared to other subjects, may have suffered particularly from austerity measures. In some countries, the share of private schooling has increased. What has happened to the quality of science education is not clear. In general, little is known about how many young people receive what kind of science education, under what conditions, and at what cost. This is a serious oversight where resources are limited, participation is uneven, and where the supply of science and technology school leavers does not correspond to the demand from higher education and the labour market.

To overcome the lack of information, the International Institute for Educational Planning conducted a special survey in 1990/1991. Teams were selected in a dozen developing countries and charged with drafting a monograph about the organization and condition of science education. Twelve countries were selected – four African countries (Senegal, Burkina Faso, Botswana and Kenya), three Latin-American countries (Argentina, Chile and Mexico), two Arab countries (Morocco and Jordan) and three countries from the Asian and Pacific region (Korea, Papua New Guinea and Thailand). This information was later complemented, for the purposes of comparison, by studies on two developed countries – France and Japan. The latter, which were prepared on the basis of published documents, were not as comprehensive as the preceding monographs. However, they did provide some interesting insights. Descriptions of learning conditions and teaching practices require more detailed studies at school level – this is set out in two special case studies on Malaysia and Morocco, where empirical investigation was carried out in selected schools.

The countries selected covered a variety of situations in terms of development level and range of economic activities, both of which influenced the type of workforce required and the resources available for investment. They also covered different types of secondary school organization and objectives for science education. The information collected focused on participation in science education (*Section 1*), the time assigned to the study of science at different levels of the education system and the importance attached to these subjects in the overall curriculum (*Section 2*), the conditions of teaching and learning (*Section 3*), teaching methods (*Section 4*), the costs of science education (*Section 5*), achievements in science and how much pupils had learnt in science (*Section 6*), and the destination of school leavers (*Section 7*).

Section 1: Participation of students in secondary science education

Participation in science education depends on the proportion of the age cohort enrolled beyond the primary level. Secondary education is usually the first level where pupils learnt science as a separate subject in a systematic way. Historically, this was true in most countries. Primary-level science is nowadays found in many school systems and is of growing importance in providing a foundation for study at higher levels. It is, however, far from universal and often very unevenly implemented. Participation also depends on the type of curriculum which is offered at secondary level and on whether or not science was taught to everybody. Finally, participation depends on the length and organization of secondary education. Science can be taught with different intensities according to the track and stream in which the pupil was enrolled. It was, therefore, deemed to be interesting to analyze not only the proportion of pupils who were exposed to some science education, but also the proportion which specialized in science at different levels in secondary schooling.

1. Organization and duration of secondary education in different
    countries

   In most school systems, secondary education covers the ages between
about 11 and 18 years. Lower secondary – the first two to four years – is
sometimes included in the definition of the basic education cycle. Upper
secondary is usually offered to children between about 14 to 18 years old.
There is much variation in duration, organization, division in cycles, and
types of streams offered within secondary schooling. Secondary education
is, however, widely given three functions – first, to broaden and reinforce
the knowledge acquired at primary level; second, to prepare students to
enter the labour market by providing them with the necessary skills and
qualifications to obtain a job or to undergo vocational training; and, third,
to deepen previous knowledge and select those who are going to continue
on to higher studies. Secondary schooling is a subject of reform in a large
number of countries. As more students were enrolled, external conditions
changed and aspirations took on new forms, curricula needed redesigning
and what was taught – and how – became an object of special concern.
   In the countries covered by the project, the duration of secondary
schooling varied from four to seven years (*Table 2.1*). In 13 countries
reported here out of 15, secondary education was divided into two cycles
– lower (or junior secondary), sometimes called middle-school education,
which lasted for three to four years, and upper or senior secondary cycle
which lasted for two to three years and, in a very few cases, four years. In
one country, Malaysia, secondary education included three cycles. The
last one (Form 6) existed parallel with some post-secondary college
courses. This cycle prepared students for entry to higher education. In
many countries it was considered post-secondary. Kenya did not have
separate secondary cycles; basic education lasted for eight years and the
whole of secondary for only four. Chile had extended its basic education
to eight years, but it still divided its four-year secondary education into
two cycles.
   Kenya and Chile were not the only countries to have incorporated
lower secondary years into a basic education cycle available to all. A
number of other countries had also made lower secondary an integral part
of an eight to nine-year basic education cycle. In Morocco, lower
secondary education, which lasted four years, had been shortened by one
year and had become the last cycle of a nine-year basic education. In
Jordan, the six-year primary education and the four-year lower secondary
constituted the basic education programme. Botswana had shortened its
lower secondary to two years which, combined with its seven-year
primary education, provided a nine-year basic education cycle. Thailand

intended to make lower secondary education compulsory as from 1995 with a view to creating a nine-year basic education programme. Argentina had removed all restrictions on entry to lower secondary and was currently introducing an eight-year basic education[1]. Malaysia allowed unrestricted access to the first nine years of education. In all of these latter cases, however, lower secondary education continued to be delivered in secondary schools with the same facilities and teachers as before. Fairly universally, lower secondary was being unified and tracking into separate vocational or technical schools was being pushed back to the end of the first cycle. Burkina Faso, Argentina, Mexico and France were the only countries in the study which continued to offer vocational streams at lower secondary level. Argentina was in the process of gradually unifying the curriculum offered in lower secondary, as was France, which was gradually reducing the size, after two years of lower secondary education, of its pre-vocational courses for academically weaker students. In some countries, out-of-school training programmes existed for primary school leavers and drop-outs. As a result of these two trends, an extended and unified basic science education was being provided to an increasing proportion of children.

At upper secondary level, the students enrolled did not generally receive the same amount and type of science education. The trend in all countries covered by the project had been to retain diversified curricula tracks. Many countries were keeping a vocational or technical stream in parallel to their academic track, and some had both a technical or technological track alongside a terminal vocational track (France, Japan, Malaysia, Mexico). The proportions of students involved varied considerably. Several countries were opting for strengthening their technical and vocational track, wanting to put it on an equal footing with academic education; this was the case in Jordan, Korea, Thailand and France. In terms of diversification at upper secondary level, Mexico represented an extreme case. There were 10 different types of upper secondary schools – some offered a general secondary curriculum, others, a technological programme combining an academic programme with specific skill preparation, and some were purely vocational; some were run by the federal government, some by the states, some by the universities, and others by private bodies. Most of the academic schools which offered a strong science content were run by the universities. As every institution

---

1. Since the studies were conducted, Botswana has planned a reform to lengthen lower secondary again to three years. Argentina has adopted an nine-year basic education, transforming its system from 7-5 to 6-3-3.

more or less defined its own programme, some 200 curricula existed. In this case, it was very difficult to assess how many pupils were learning how much science.

Despite differences in the subdivision of educational cycles, the total number of years needed to reach the end of secondary education without repetition varied very little. It amounted to 12 years in almost all countries except Senegal, Burkina Faso and Malaysia, where it took 13 years (11 years in Malaysia, if Form 6 was not included). Some patterns were more influenced by the French tradition and others by the English.

2. Participation in science education

Participation in science at lower secondary level

At lower secondary level, all the countries covered by this project included mathematics and at least one science subject in their curricula. The last countries to maintain programme diversification at lower secondary level, had included the sciences in the curriculum of all of their tracks. The participation rate in science at that level was therefore equivalent to the enrolment ratio at lower secondary level (see *Tables 2.1* and *2.2*).

Participation varied considerably. As could be expected, the lowest participation rates in science, as measured by gross enrolment rates, were found in the least-developed countries – 8 per cent in Burkina Faso and 21.4 per cent in Senegal. The figure was also low in Papua New Guinea (16 per cent), a country with a similar level of development to Morocco (45.2 per cent),[2] Botswana (55 per cent) or Thailand (34.4 per cent). By integrating its lower secondary into basic education, Kenya probably succeeded in providing some science education to more children at this level at a lower cost.

2.   Morocco, however, had not succeeded in raising the enrolment ratio at primary level. In this country, access to lower secondary was open well before universal primary education was achieved, which was not the case in Papua New Guinea.

Table 2.1   Education system coverage and flows between different levels (1990-1991)

Countries	Education system structure (1)	Gross enrolment rates (%)				Transition rates (%)			Lower second. Unified (9)	Upper second. Unified (10)
		Primary (2)	Lower second. (3)	Upper second. (4)	Post-second. (5)	Primary/ second. (6)	Lower sec/upper sec (7)	Second./ post-sec (8)		
Burkina Faso	6 (#4#3#)	37.0	8.0		2.4	0.7	32.3 G 10.1 T	28.2 G 10.1 T	18.4	G T	G V
Senegal	6 (#4#3#)	58.0	21.4		10.1	2.9	24.9	62.6 G 9.7 T	39.1	Yes	G T&V
Kenya	8 (#4#)	95.0		29.0		2.1	41.2	na	6.2	G (VT)	G VT
Botswana	7 (#2#3#)	110.0	55.0		19.0	3.0	68.0	40.0	55.0	Yes	G (VT)
Morocco	6 (3#3#)	62.4	45.2		19.4	9.9	87.2	36.5 G 1.2 T	64.6	Yes	G T (VT)
Jordan	6 (4#2#)	97.0	90.0		57.0	21.7	98.0	48.7 G 16.0 T 2.3 V	40.3	Yes	G T V
Papua N. G.	6 (4#2#)	72.0	16.0		2.0	2.0	37.0	10.5 G 25.0 VT	88.0	Yes	G T&V
Thailand	6 (3-3)#	86.5	34.4		22.5	15.7	44.7	45.0 G 34.9 VT	90.3	Yes	G V
Malaysia	6 (3#2#2)	99.8	83.0		49.1	18.9 PS 2.8 U	88.4	68.0 9.7 T	34.0	Yes	G T V
Korea	6 (3-3)#	107.0	98.0		86.0	37.7	98.0	63.0 G 33.0 VT	47.0	Yes	G V
Chile	8 (2-2)#	98.0		75.7		20.6	94.0	68.2 G 25.8 T	81.0	Yes	G T&V
Mexico	6 (3-2 to 4)	110	57.8		29.8	14.1	81.1 49.8 G 31.3 VT	74.0 59.0 GT 15.0 TV	106.0	Yes	G V T TTC
Argentina	7 (5)	111.9		71.0		39.9	82.5	91.7 G 83.6 T	120.0	G T Agr.	G C T
France	5 (4-3#)	109.0	99.2		84.9	39.6	93.8	59.8 GT 21.9 V	68.3	Partly	G T V
Japan	6 (3-3)	101.0	99.9		94.1	28.7	100.0	94.1 71.8 G 28.2 TV	53.7	Yes	G T V

na = not applicable.
# = Examination.
G = General (academic) education.
T = Technical secondary education.
C = Commercial secondary education.
TTC = Teacher training colleges.
V = Vocational education.
VT = Vocational training
U = University
PS = Post-secondary
Source: Caillods, Göttelmann-Duret, 1991, and country monographs.

Table 2.2   Proportion of pupils enrolled in general (academic) secondary
education and attending science courses (1990-1991)

Countries	(1) Proportion of secondary enrolment in general stream		(2) Content diversification within general secondary		(3) Science taught		(4) Participation rate in science education				(5) General upper secondary % pupils specializing in science
	Lower second.	Upper second.	Upper secondary		Lower second.	Upper second. Arts stream	Lower secondary		Upper secondary		Total science streams
Burkina Faso	95.0	79.0	Yes	4 streams	Yes	Partial	14-17 years	8.0	18-20 years	1.9	63.2
Senegal	100.0	84.7	Yes	3 streams	Yes	Partial	13-16 years	21.4	17-19 years	8.8	51.0
Kenya		na	Weak		nr	Yes	14-17 years			29.0	nr
Botswana	100.0	na	Yes	Options	Yes	Yes	14-15 years	55.0	16-18 years	17.0	42.0
Morocco	100.0	94.4	Yes	4 streams	Yes	Partial	12-15 years	45.2	16-18 years	17.6	52.9
Jordan	100.0	72.7	Yes	2 streams	Yes	Yes	12-15 years	90.0	16-17 years	41.4	41.6
Papua N. G.	100.0	25.0	-	-	Yes	Partial	12-15 years	16.0	16.7 years	0.5	na
Thailand	100.0	59.5	Yes	3 streams	Yes	Yes	12-14 years	34.4	15-17 years	13.4	56.1
Malaysia	100.0	91.6	Yes	2 streams	Yes	Yes	12-14 years	83.0	15-17 years	49.1	25.6
Korea	100.0	62.3	Yes	2 streams	Yes	Yes	12-14 years	98.0	15-17 years	53.5	46.0
Chile		72.5	Yes	Options	Yes	Yes	14-17 years			*75.7/ 72.5	na
Argentina	39.0	44.2	Yes	12 streams	Yes	Yes	13-17 years			*71.0/ 44.2	17.8
Mexico	61.4	60.3	Yes	10 types of schools	Yes	Yes	12-14 years	57.8	15-17 years	23.8	na
France	92.0	45.9	Yes	5 streams	Yes	Partial	11-14 years	99.2	15-17 years	39.7/ *55.0	46.4
Japan	100.0	72.5	Options Group levels		Yes	Yes	12-14 years	99.9	15-17 years	67.7	na

nr =   not relevant: na = not available; (1) Proportion of secondary pupils enrolled in general academic education (i.e. not in technical or vocational education); (4) Proportion of the age group studying science. At lower or at upper secondary, this was generally computed as the proportion of the age group enrolled in general (academic) secondary. Where there was clear evidence that pupils in technical streams studied science, the total enrolment ratio is also given*. (5) Proportion of pupils enrolled in a science stream or taking science options.

Source: Caillods, Göttelmann-Duret, 1991, and country monographs.

The participation rate for the 12 to 14 year-old in science education was very high in Japan, France, Korea and Jordan, reflecting their high gross enrolment rates at lower secondary level. It increased progressively together with the level of economic development but also depended on the policy for access to lower secondary, and on the level of drop-outs during primary education. Considering its level of development, Mexico had a rather low participation rate in secondary and, hence, in science education. This did not seem to be due to selection at secondary entry but rather to continuous drop out throughout primary and lower secondary.

Participation in science at upper secondary level

In most of the countries covered by this project, pupils enrolled in *academic* upper secondary education studied mathematics and at least one science subject (general science or natural science), if only for a few periods a week and during one or two years of the programme only (see *Section 2* of this chapter). There was just one exception – Papua New Guinea – where certain pupils could choose not to take a science option at secondary level.

Students enrolled in industrial technical or vocational education generally studied mathematics and one or several science subjects; this was not the case however for students enrolled in commerce streams or in many vocational streams – they might study mathematics but not necessarily science. In the absence of information on the curriculum and number of pupils enrolled in each section of technical and vocational education, enrolments in these streams were not included when estimating the participation rate in science education at upper secondary level. Thus, this participation rate has been estimated, with one exception, as equivalent to the enrolment rate in *academic* upper secondary education (*Table 2.2*). These rates are gross enrolment rates; they may be overestimating the participation in countries which had high repetition rates and varying numbers of over-age students. Hence, they have to be interpreted carefully!

The resulting proportion of the age group receiving science education varied in upper secondary even more than at lower secondary level (see *Table 2.2*). It was low in French-speaking African countries (1.9 per cent in Burkina Faso and 8.8 per cent in Senegal) as well as in Papua New Guinea (0.5 per cent). Botswana, Morocco and especially Kenya had a much higher proportion of their age group studying science at upper secondary level. Kenya, however, incorporated at this level a younger age group than many other countries. Thailand, and even more so Mexico, had, on the contrary, a relatively low participation rate in science for the relevant age group considering their respective level of development and industrialization. In these three countries, however, a fairly high proportion of the age groups were enrolled in vocational education, so the real rate of participation in science was probably underestimated. The highest rates of participation in science education at upper secondary level were to be found in Japan, Chile, Korea and France. Jordan and Malaysia also had a fairly high rate of participation.

Specialization

In almost all countries, curricula specialization occurred even if the degree of specialization varied considerably. In the Francophone countries (France, Morocco, Senegal and Burkina Faso), pupils were grouped into streams that prepared them for different *baccalauréat* series. Each stream had a corresponding pre-established menu of subjects and there were two scientific streams. In countries with an Anglophone tradition there might be less restriction on the subjects, and students elected from a range of options which were offered in their school. They were nevertheless frequently grouped in science and art streams, and this determined their compulsory subjects and the range of possible options. Elsewhere, specialization might result from the type of schools the students attended rather than through the choice of options or streams; this was the case in Mexico and, to a certain extent, in Japan.

Comparing the percentage of pupils who specialized in science education and assessing the proportion of the relevant age group which was thus specialized would have been interesting, and revealing of underlying policy. This was not easy, however, since not all countries collected this information. It was particularly difficult to unravel in cases where specializations were not reflected in streams so much as in option choices and these led to different degrees of specialization, as in Papua New Guinea, Chile and Japan. It was then necessary to rely on examination data at the end of secondary – and even these were possibly ambiguous, since those taking more than one science subject were not usually identified as a separate group. Where there was no such examination, and where admission to higher education depended on an entrance examination or an aptitude test, the information on specialization was often just not available.

The information available on the different countries appears in *Table 2.2*. Analysis of these data show that in many of the developing countries included in the project, the proportion of pupils specializing in science education at upper secondary level is high – above 40 per cent. In four countries – Burkina Faso, Senegal, Morocco and Thailand – more than half of the students enrolled in academic secondary are specializing in science. Three groups of countries seem to emerge:

In the first group of countries, a very high proportion of students specialize in science at secondary level. This proportion and the actual numbers are in fact higher at secondary than at higher education level; this was the case in Senegal and Burkina Faso (see *Figure 2.1*). Students who could not enter science courses at post-secondary level had to find a place in other courses or join the labour market.

Figure 2.1. Percentage of students specializing in science at upper (general) secondary and post-secondary levels (1990)

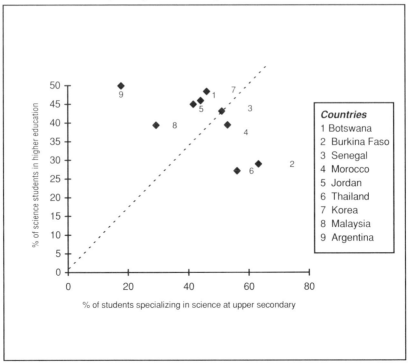

Source: Caillods, Göttelmann-Duret, 1991, and country monographs.

In a second group of countries, the proportion of students who specialize in secondary education is high, and more or less equivalent to the proportion of students who specialize in this field in post-secondary education.

In a third group of countries, the proportion of students specializing in science at secondary level is low – lower than at higher education level. Malaysia was one example, as were Chile and Argentina. The mismatch was such that it could create problems for higher education institutions which may lack sufficient numbers of suitable candidates; in Chile, for example, the number of students applying for university or higher education courses in science and engineering declined drastically between 1980 and 1991 (in biology by 64 per cent, in physics by 85 per cent and in chemistry by 88 per cent) (Schakmann, Zepeda and Toro, 1992). In order to address the shortage of applicants for post-secondary science-based programmes, higher education institutions in Chile (and similarly in Argentina) have had to recruit students who were not specialized in science and special courses were offered to them in order to reach standards necessary for higher education. This is similar to arrangements made in Malaysia for some pre-university programmes to upgrade students with a low level of specialization to prepare them for science-based courses. Mexican universities ran their own upper secondary schools (preparatoria) for similar reasons.

Section 2: Curriculum organization

The secondary science curricula which are being institutionalized in different countries share a certain number of common objectives and "converge towards a largely common body of science content which constitutes the science taught during the 12 or 13 years of schooling" (Rosier and Keeves, 1991). The comparative analysis of the patterns of the officially prescribed or recommended secondary science curricula conducted for this project revealed that countries differed in various aspects (see e.g. Postlethwaite and Wiley, 1992; IAEP/ETS, 1992b). The differences pertained in particular to the more or less integrated form with which science was to be provided at lower secondary level, the curriculum time set aside for maths and science at different levels and at upper secondary level, and to the extent and forms of specialization. These points will be looked at separately at lower and upper secondary levels.

The discussion below is based on the official curriculum, even though there may be significant differences between the official curriculum and the one implemented at classroom level. It should be noted in this respect that in the vast majority of countries, the science curriculum to be taught through primary and lower secondary grades was specified at national level. But while in some countries the same science curriculum was imposed on all schools in a rather restrictive way, in other cases, regional authorities and schools were given wide margins for manoeuvre in interpreting and implementing the official guidelines. Such was the case in Argentina, Chile and Mexico where only very general non-directive curriculum guidelines or a specific core curriculum for Federal State Schools only (Mexico) were issued at national level[3]. This left a great deal of leeway to regional authorities and individual schools in the determination of the science curricula actually offered to pupils.

1. Organization of science education at lower secondary level

In many systems, the teaching of science started before pupils entered secondary education, often in primary grade 3 or 4 when it is assumed basic literacy and numeracy have been established. At this level, science was taught in combination with other curriculum fields (the case of 'life experience' in Thailand), as an integrated subject (e.g. in Kenya) or in the form of a combination of two or more science subjects (the case of 'natural sciences' in Morocco, combining, for example, biology and geography). With respect to science achievements at later stages of schooling, a policy of introducing science early seems to be an effective one; the international IEA study on science achievement found that science achievement of 14-year-old secondary students was very closely associated with science achievement at primary level (Postlethwaite and Wiley, 1992).

At lower secondary level, policy options differed among the various countries as to the amount of curriculum time set aside for mathematics and science, subject combinations and organizational patterns through which basic science education was provided. As shown in *Figure 2.2,* the average number of hours of science and mathematics instruction that lower secondary students received per year varied significantly – from

---

3. The pros and cons of each of the two options are clear – disparities among schools as regards the science content imparted to students tends to be rather limited in systems imposing the same science curriculum on all schools (and which closely control its implementation). Disparities are probably greater in less 'centralized' systems, but the latter may be in a better position to adapt the organization, content and methods of science teaching and learning to the specific conditions and needs of pupils at local or school levels.

180 in Jordan to 432 in Botswana; in Thailand and Japan – two countries in which the 14-year-old pupils showed on average good science achievement levels in the IEA study mentioned above – lower secondary students received around 220 hours of science and mathematics instruction per year (according to the official curriculum).

Lower secondary students received a more or less 'mathematics-biased' scientific instruction depending on the country. In countries following the French model, like Senegal and Morocco, lower secondary students spent almost twice as much time on mathematics as on science subjects; the 'mathematics bias' also existed – although to a lesser extent – in Argentina and Chile. In other countries (e.g. Botswana, Kenya, Thailand), the number of science lessons at lower secondary level was significantly higher than that set aside for mathematics. According to the national curriculum valid at the beginning of the 1990s, pupils in Thailand could even opt out of mathematics completely from the last grade of lower secondary education.

The impact that these different emphases on mathematics may have in terms of the further career options of lower secondary school leavers in vocational training programmes or on the labour market has not been studied. With regard to selection into science programmes of the academic type at upper secondary level, the importance of a good performance level in mathematics seems, however, to be widely agreed. Even in some of the countries in which lower secondary students spent less time on mathematics than on science, e.g. Thailand, it was students' scores in mathematics rather than science that determined their access to academic upper secondary education in general and to science streams or programmes at this level in particular.

Regarding the organizational patterns of science provision, a number of countries opted for providing basic science education at lower secondary (as well as science for non-science students at upper secondary level) as one subject, without divisions into biology, physics and chemistry. Thus Botswana, Chile, Kenya, Korea, Japan, Jordan, Malaysia, Mexico, Nigeria and Thailand, mostly countries following the English or American model, offered 'integrated science' or 'general science' courses in which science was taught as a whole. Many of these integrated lower secondary science courses had their origins in, or had been clearly influenced by, integrated programmes designed in developed countries.

Figure 2.2. Average number of teaching hours spent on science and mathematics*
  (1989/1990)

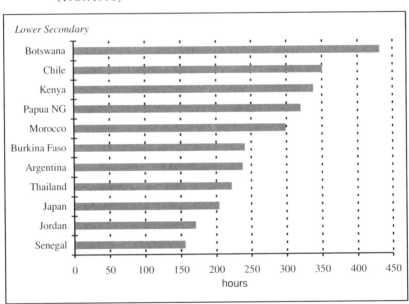

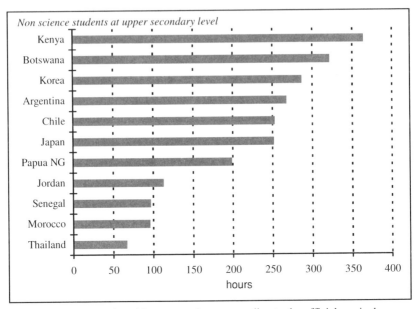

* Minimum number of teaching contact hours according to the official curriculum.

Source: Country monographs, 1989-1990.

France and the Francophone countries included in the project
– Burkina Faso, Morocco, Senegal – had not adopted this option, but at
lower secondary level biology and geology were combined in a 'natural
science' course and physics and chemistry in a 'physical science' course.
In some Latin-American countries – Chile, for example – biology, physics
and chemistry continued, on the other hand, to be taught as separate
subjects at all levels and for all students in the compulsory programme at
lower secondary level.

The implementation of integrated science courses encountered
difficulties. In some countries, these courses were actually not taught by
one teacher, as was generally recommended, but subdivided between
several teachers to reflect disciplinary boundaries. The main obstacles
mentioned in IIEP's international survey data were: insufficient
preparation of teachers for the task; reluctance of teachers to go beyond
the traditionally established boundaries; lack of appropriate teaching and
learning materials; and conflict with established timetables and school
organization. In many countries, teacher-training courses maintained
separate disciplinary-based teaching of science for prospective science
teachers.

Irrespective of their more or less integrated form of provision, lower
secondary science programmes showed a common trend towards linking
up science with children's everyday life and environment, dealing with
health, nutrition, natural habitat, and sometimes agriculture. Furthermore,
as regards the approaches and methods of science teaching and learning,
official curriculum documents were found almost everywhere to attribute
a central role to discovery learning (including laboratory experiments but
also other approaches using observation, 'guided discovery' and practical
activities) as well as to child-centred teaching and learning approaches.
The implementation of these methods requires that science education
means more active individual learning and group work and fewer 'chalk-
and-talk' lessons. The implementation of these objectives seemed to be
impeded by the large amount of curriculum content and number of topics
that teachers and students had to cover (in science and in other subjects)
in order to succeed in the main high-stake examinations.

2.   Science curricula and specialization at upper secondary level

At upper secondary level all countries included in the research
offered some possibility of specializing in science. The more advanced
programmes in these subjects (provided to 'science students' preparing
for entry into science-related courses at the tertiary level) were similar
– they commonly offered a combination of biology, chemistry and physics

taught as separate subjects. However, the originators of the upper secondary curricula of various countries expressed rather different views as to the necessary degree and profile of specialization of 'science' at upper secondary level. In particular, the minimum amount of science education to be taken throughout the cycle, the time that specialized students were meant to spend on science and mathematics and the difference between the programmes for 'science' and 'non-science' students varied significantly from one country to another.

Curriculum patterns for science students

As regards the *curriculum patterns for students specializing in science*, three different approaches or 'models' can be distinguished (see *Figure 2.3*). The first model – found in Kenya and Chile, for example – leaves very limited scope for specialization since the minimum number of mathematics and science lessons that students specializing in science had to take was only very slightly higher than the compulsory programme for non-specialized students in those subjects. At the other extreme, certain countries offered students a large scope for specialization – at least on paper – by allowing them to select the science and mathematics courses that they would like to follow; this could lead to high degrees of specialization in the sense that students specializing in science were allowed to spend a very limited share of curriculum time on subjects other than science and their programme of specialization could differ a lot from those of non-science students (Model 2). This approach existed in particular in countries influenced by the British or the American system of education (Botswana, Japan, Korea, Malaysia, Thailand, etc.). Thailand was (according to the curriculum applied in the early 1990s) an extreme case, to the extent that science students could even opt out of mathematics if they so wished; in Japan and Korea, on the other hand, the range of compulsory subjects to be taken by students specializing in science was rather broad. The 'mixed' approach which was adopted in France (Model 3) and by countries influenced by the French system is situated between these two models. This approach offered upper secondary students the possibility of choosing one out of several science or non-science streams. One year later, there was a choice between two different science programmes (one focused on mathematics and physics; the other on biology and chemistry). For those enrolling in science streams, the share of curriculum time set aside for mathematics and science did not exceed 50 per cent, i.e. ample room was given to non-science subjects (see *Figure 2.3*).

Figure 2.3. The time shares of science and mathematics in different
systems of secondary curriculum organization

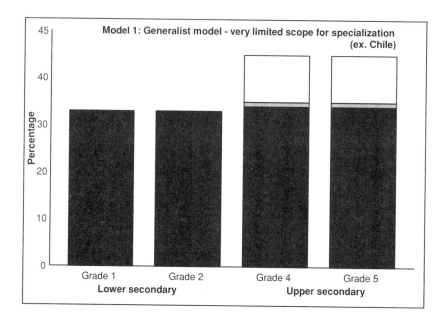

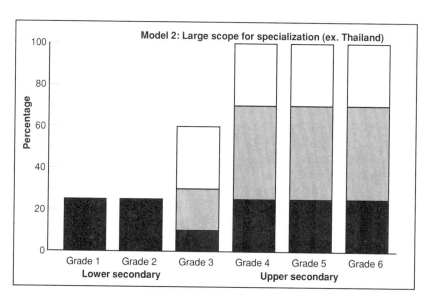

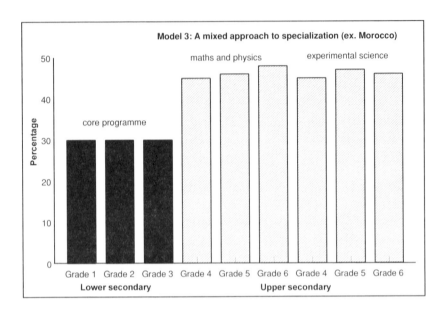

Model 3: A mixed approach to specialization (ex. Morocco)

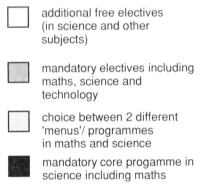

additional free electives
(in science and other
subjects)

mandatory electives including
maths, science and
technology

choice between 2 different
'menus'/ programmes
in maths and science

mandatory core progamme in
science including maths

Source: Caillods, F.; Göttelmann-Duret, G. 1991. Science provision in academic
              secondary schools, organization and condition, Paris, IIEP mimeo.

Model 2 – if it leads to high levels of specialization of secondary science graduates – may impede flexible adjustments to the training needs at post-secondary level and the requirements of the labour market. Countries following policies of low specialization (the first model), on the other hand, are running the risk of a shortage of secondary graduates who are sufficiently interested and adequately trained in science to take up further studies or training in related areas. This was apparently the case in Chile as mentioned earlier in this chapter.

Curriculum options for non-science students

There are three types of policy regarding the curriculum options for non-science students: in the *first* type of policy, students were allowed to drop certain science subjects or mathematics completely from their entry into specialized programmes at upper secondary level; thus, it was possible for senior secondary students in Botswana to opt out of science after the first year of the upper secondary cycle and for Thai students to drop mathematics completely over the whole duration of this cycle. In the *second* type of policy (notably in countries following the French tradition), non-science students were taught very few hours of 'natural sciences' during the upper secondary cycle (in Morocco, for example, these students received a single period of 'natural sciences' per week in the first and second year of upper secondary and none at all during the last year of this cycle). A *third* policy consisted of imposing a substantial minimum amount of science (and mathematics) lessons on non-science students until the end of the upper secondary cycle; the latter option has clearly been adopted in Japan and Korea as well as in a few other countries, e.g. Kenya and Chile. *Figure 2.3* demonstrates the significant variations among countries as regards the minimum amount of science instruction that their curricula stipulate should be imparted to all students throughout secondary education.

Countries in which the learning of science started early and in which the subject continued to be imparted *to all* until the end of secondary (albeit in differentiated forms) produced non-science secondary graduates with a substantial knowledge base in science. This may be considered as an asset with respect to both the country's development in general and the prospects for flexible adaptation of human resources to the changing needs of the labour market.

3. Selection and assessment

Access to upper secondary and post-secondary level and option choices available to students usually depended on the results of assessment carried out at the preceding level or at the stage of admission to higher levels. Modalities of assessment varied. Some countries organized external standardized examinations at the end of upper primary (as in the case of Kenya) or lower secondary cycle (Botswana, Burkina Faso, Senegal, Papua New Guinea, Malaysia); in Argentina, Korea, Mexico and Thailand there was no nationwide central regulation of access to upper secondary education, but many secondary schools, especially the most prestigious ones, ran their own entrance examinations. There has been a growing tendency to base decisions on promotion from lower to upper secondary on the results of continuous assessment or a mixture of continuous assessment and school-based examinations, as in Morocco.

In all the countries surveyed, the transition from secondary to post-secondary education was regulated by an external – national or regional – examination either at the end of secondary or at university entrance level and, as mentioned earlier, access to post-secondary science courses depended on the scores obtained at this examination in whatever subjects were appropriate. Almost everywhere a number of post-secondary science and technology courses had special admission requirements or ran their own recruitment tests alongside the mainstream system.

Access to upper secondary courses usually required more evidence of achievement in mathematics than in science. In very few cases (e.g. Jordan), student selection into upper secondary education depended as much on science as on mathematics scores. With the exception of Kenya (where the weight given to these two subjects was relatively low), mathematics and science taken together made up between 37 and 50 per cent of the overall score in admission examinations at this level. Once students had obtained admission to upper secondary, their access to different science streams or options depended on different criteria, according to the specific case considered. In countries following the English or American pattern, allocation was the result of a mixture of preference and satisfactory minimum performance in lower secondary mathematics and science; scores in other subjects were usually less important. In Francophone systems, on the other hand, access to science streams, particularly to the 'mathematics and physics' stream, required high overall performance, although mathematics and science subjects were given extra weighting.

The 'subject profile' of secondary graduates selected into post-secondary science courses also varied across countries. In countries where many specialized higher education institutions had their own admission criteria, access to post-secondary science courses tended to be based on the achievement scores obtained in the science subjects for which they applied. In other cases – especially where selection was based on a single selective examination (Secondary Leaving or University Entrance Examination) – admission to science programmes in universities or other institutions tended to depend not only on students' scores in science but also on the average scores that they had obtained in this high-stake examination.

The central issue arising here – which will be discussed in Chapter III – is the effectiveness of different modes of assessment for guiding and selecting students in a reliable, appropriate, and equitable manner.

Section 3: Teaching and learning conditions in science education

Science teaching conditions do not differ fundamentally from the conditions of secondary education as a whole. Numerous studies have demonstrated that these conditions deteriorated in the 1980s in a large number of developing countries due to the economic crisis and associated adjustment programmes. Teachers' salaries had drastically declined in real terms in many African and Latin-American countries. In some cases, textbooks and equipment had become a rarity. Since the international conference on Education for All at Jomtien in 1990, more funds had been allocated to the education sector but, as a number of monographs written in the framework of this project indicated, these funds had been directed primarily towards basic education. Some efforts had also been made to restore the quality of higher education. In the meantime, teaching and learning conditions at secondary level continued to deteriorate. Science education was likely to be the most affected, because of the greater costs involved in its organization.

In fact, the situation encountered varied widely from country to country and from region to region. It gives cause for most concern in the African countries included in the IIEP project; however, some middle-income countries had made substantial and successful efforts to improve the teaching conditions in their schools. But there was evidence of continued dissatisfaction even in the most advanced countries in IIEP's sample. This section reviews briefly the situation of the teaching and non-teaching staff, and then the situation vis-à-vis specialized equipment and facilities, textbooks and other support for teachers.

1. Availability of science teachers and turnover

The African countries in the sample as well as Papua New Guinea, still suffered from a serious shortage of mathematics and science teachers. Some had recourse to expatriate teachers – between 5 and 16 per cent of the teachers in Senegal, Burkina Faso and Papua New Guinea were expatriates. Botswana represented an extreme case – 56 per cent of all science teachers at junior secondary level and 75 per cent at senior secondary level came from abroad in 1990. The country of origin of expatriate teachers was also changing and becoming more diversified – in Botswana, again at senior secondary level, a third of the teachers came from India, a quarter from the United Kingdom and another quarter from other African countries.

Calling on expatriate teachers created a certain number of problems. First of all, these teachers were expensive, even when their salaries were paid by the country of origin (this was generally only the case for European and American citizens, and was becoming less and less common). Second, they were not always familiar with the curriculum they were supposed to be teaching. Third, they sometimes did not have mastery of the language of instruction, which could create serious problems of communication with the pupils, who themselves did not always understand the medium of instruction. Dependence on expatriates often made the use of a national language difficult. Shifting to Arabic in secondary education was made possible in Morocco only when the country stopped using expatriate teachers. Finally, expatriates often had a high rate of turnover, as most were on contract for two to three years. Serious problems could occur when teachers left in the middle of a school year because their contracts had expired. They were not necessarily replaced, or substitutes had to be found who had to familiarize themselves with the school and the curriculum in a very short period.

Since expatriate teachers were expensive, some countries either did not call upon them at all or not in sufficient number. Science and mathematics might then be taught by a teacher who had not been trained in the discipline, or not trained at all. This phenomenon is referred to in the monographs which were prepared on Senegal (Seck Fall, 1991) and Burkina Faso (Daboué, 1990) – but the real extent of the problem in these countries and elsewhere is generally unknown.

Turnover and dropout was another major problem among science teachers, independently of the presence of expatriate teachers. The situation in this respect however very much depended on the condition of the local labour market. It was a serious problem in countries like Thailand and Botswana, where the labour market was developing and where people with a good science background in mathematics and science were in short supply in the private sector. It was a problem also, but to a lesser degree, in Latin-American countries (Mexico and Chile) because of the drastic decline in teachers' salaries. But it had paradoxically ceased to be a problem in many African and Arab countries where alternative jobs were scarce and less secure. Senegalese and Jordanian teachers who used to go abroad to teach in countries offering higher salaries (such as Gabon, Algeria or the Gulf countries) have returned home. Likewise, in countries experiencing a public sector hiring freeze and private sector slowdown, science teachers no longer made attempts to leave the profession.

Other countries (Morocco, for example, and – curiously – Korea) had trained so many science students at higher education level that their graduates suffered from unemployment; hence, they no longer hesitated to take up a job which was viewed as stable even if the salary was thought to be unsatisfactory.

2. Qualification of science teachers

Some countries, such as Kenya, Senegal and Jordan, were suffering from a serious shortage of qualified science teachers. In Kenya, in particular, over one-third of the teachers in state secondary schools were untrained. This percentage was actually higher in the assisted schools and the former Harambee schools than in the fully government-maintained schools. It was due to the rapid expansion of schools and to the fact that the country no longer recruited expatriate teachers. This high proportion of untrained teachers applied to all subjects but the number of unfilled vacancies was much higher in science and technology subjects.

In other countries, the proportion of qualified teachers was fairly high. This has often been achieved by certifying lower secondary teachers on sub-degree training courses, while requiring upper secondary school teachers to be trained at university-graduate level. The statistics on the proportion of qualified teachers in *Table 2.3* may be misleading if the teachers who were trained to perform at lower secondary level actually taught at upper secondary level (as was often the case) or if they did not teach the subjects for which they were trained. Surpluses of some specializations can exist with shortages of others; in many countries, some science was taught by specialists trained in other science options. Most commonly, biological science teachers substituted for physical science teachers, who were usually in shortest supply.

Table 2.3   Staffing conditions in secondary science education (1990)

| Countries | Number of pupils per class | | | | Class split in groups for laboratory work | Proportion of science teachers who are: | | Shortage of science teachers | Number per 100 teachers: | |
| | Lower secondary | | Upper secondary | | | | | | | |
	Public	Private	Public	Private		Qualified	Expatriate		Labora-tory assistant	Science inspector
Burkina Faso	64.2	60	21/55*	25/57*	-	100.0	4.6	Yes	ε	1.1
Senegal	50	45	38/49*	21/30*	Yes	89.0	7.5	Yes	ε	1.6
Kenya			36	55	No	61.1	0.0	Yes	ε	0.5
Botswana	40-45*		30-35*		No	63.0	56.0-75.0*	Yes	ε	na
Morocco	28	19.8	25/32*	35/28*	Yes	98.0	0.5	No	10.0	1.6
Jordan	28	22	30	21	No	66.0/33.0	0.0	No	1.1	1.1
Papua N. G.	40-35**		20-25**		Yes	na	16.0	Yes	ε	na
Thailand	41	39	32	29	No	100.0	0.0	No	ε	2.6 3.2
Korea	50.2		53.6		No	100.0	0.0	No	Insuff.	Suffic.
Chile	Varying				No	100.0	0.0	No	none	Insuffic.
France	24.3	24.4	31.5	25.7	Yes	100.0	0.0	No	10.0	na

*   according to the stream or the science subject taught – the maths/physics streams have lower class sizes than natural science.
** recommended class size.
ε   negligeable.
na not available.

Source: Country monographs, 1989-1990.

The Malaysian data illustrate some aspects of this. Overall, the number of science teachers was sufficient in 1990. However, all did not teach science and about 30 per cent of all those teaching science subjects were trained in other subjects, most frequently mathematics. The mismatch between the options that teachers were trained in and the subjects they actually taught was worst in general science and physics at upper secondary. At lower secondary, there seemed to be a lot of substitution, with teachers trained in one or more science subject teaching across the integrated curriculum (Sharifah Maimunah and Lewin, 1993). To the extent that there is substitution, judgements have to be made about how far this is reasonable – physics specialists can probably teach lower secondary integrated courses, but can or should mathematics specialists teach upper secondary biology?

The training of polyvalent teachers makes their deployment easier. Morocco had only two broad types of science teachers – physical science and natural science teachers. As a result of this policy and because output of teacher training had been high, virtually all teachers were teaching the subject for which they were trained. If anything, there was a surplus of science teachers.

In Mexico, Argentina and Chile, the problem was different. In general, these countries had trained enough teachers and most of them were qualified. In the 1980s, however, teachers' salaries declined dramatically in real terms. In order to maintain their standard of living, most teachers then taught in more than one school (day schools, night schools, state schools and private ones). In Chile, it was estimated that 40 per cent of teachers taught in another school, 53 per cent taught some six to eight classes, and a third taught more than nine classes. One teacher out of four taught as many as 360 pupils (Schakmann, Zepeda and Toro, 1992). In Argentina, the term 'taxi-teacher' was used to describe teachers who moved from one school to another. Under these conditions, teaching preparation was likely to be hurried, homework sporadic, and practical work superficial.

3. Teacher support

School effectiveness might have different dimensions in different countries. In Mexico, schools judged to be effective appeared to have the following characteristics (Rojo and Martinez, 1991):

- a well-trained teaching staff teaching in only one school;
- a relatively stable teaching staff which helped teachers identify with the school, and ensured continuity between classes and cycles;
- the existence of many opportunities for interchange between teachers, with the school authority, and with the parents in the community;
- a relatively low ratio of student to teachers and a high ratio of educational resources to students;
- the fostering of extracurricular activities.

These characteristics were recognized as desirable in many of the other systems we studied. Providing an adequate support to teachers through, among other things, enhanced opportunities for professional development, and better access to resources is another condition for improved classroom teaching. Several support mechanisms are evident from the research.

Support from principals, other colleagues and supervisors

Head teachers' support was generally recognized as an essential element in fostering teachers' development. They are usually responsible for the allocation of resources and can mobilize support from within the school, from the community, and from higher authorities. They also have line management responsibilities for staff and are therefore in a good position to provide encouragement and leadership.

It was quite common to find that most head teachers did not have a science background. This was the case in Malaysia and in Morocco (where respectively only 17.7 per cent and 5 per cent of the head teachers were former mathematics or science teachers, the majority coming from a mathematics background), as well as in many other countries. The role of the principal was found to be more administrative, with little direct involvement and concern for classroom activity. A high proportion of heads in the Moroccan schools did not attend the meetings organized to discuss subject-based issues in science. In Malaysia, some heads in the schools where case studies were conducted were frequently absent on other legitimate business, thus making day-to-day decision-making on science in the school problematic. Some of the data suggest the head might be perceived more as someone who could foresee difficulties rather than as an inspirational leader in science subjects. This does not mean that these head teachers cannot exercise leadership in science education provided they receive appropriate training and are motivated. It does suggest that senior science teachers should have a role to play in the management process, especially where heads have no special science expertise.

Senior science teachers did not exist in all systems as a recognized category in the hierarchy. In Malaysia, their status had recently been improved, but problems still remained. In some cases, relatively junior staff, who might be more qualified than senior staff, were promoted rapidly and might have had difficulty managing those more senior in years. In Morocco, some teachers were supposed to be appointed to the role of subject co-ordinator at the school level, but in practice they either did not exist or had very limited responsibilities. In every Moroccan school there was a teacher responsible for the laboratories. His/her role was limited to ordering equipment and consumables. The fact that no real financial or status advantage was attached to the position had led, in some cases, to the nomination of very junior teachers to that position, which was creating problems.[4]

The provision of *science inspectors* was inadequate in all the countries studied (except Morocco). Often one inspector covered several hundred teachers and many more schools than could be visited in a year. In many African and Latin-American countries, education inspectors no longer visited schools, or they only visited the ones in or near main towns, since there was a lack of means of transportation or the money to pay for it. The image of inspectors was not always very positive. As the title 'inspector' suggests, they were seen in some countries – particularly in the Francophone ones – as someone who controlled rather than as someone who advised. Their main duty was often to recommend fully qualified status to teachers-in-training or to visit and inspect a teacher due for promotion rather than to help improve the quality of science education.

In Malaysia, several divisions of the Ministry of Education were involved in the supervision of science education in secondary schools. Among the case-study schools there appeared to be very little direct supervision of science by inspectors. Visits were more often block inspections, not specific to science education, or confined to special events. It was quite common for Malaysian teachers to be observed in the school by other staff members, though this was not the practice in many other countries. External visits to the science departments were far less frequent, with many schools receiving none over several years. In Morocco, on the other hand, science inspectors visited secondary schools regularly; however, a majority of new teachers – who would appear to need special support – reported that they had not been visited during their first years of service.

Textbooks

There were various issues related to textbooks – including their availability, which was closely related to their cost, and their quality and relevance.

---

4.   In Morocco, teachers responsible for the laboratories were supposedly asked to teach fewer periods than other teachers. With the situation of over-supply of science teachers, the teaching load of all teachers had been reduced and this relative advantage had disappeared. In 10 per cent of the cases surveyed, newly appointed teachers had been asked to do this task.

In some countries, all science textbooks were imported. This was still
the case in small countries, particularly in Africa (Burkina Faso, Senegal
for upper secondary schools) and Papua New Guinea in the Pacific. In
view of their costs, and distribution problems, availability was often a
problem at school level. In other countries, textbooks were designed and
produced locally, sometimes by the Ministry of Education (Jordan,
Morocco), by an Institute under the control of the Ministry of Education
(Korea, Thailand), or by private publishers (Mexico and Chile, and
Malaysia).

Information on textbook availability is not usually collected and
requires a special survey. In principle, school textbooks were lent to lower
secondary pupils in Papua New Guinea, Jordan and Korea. Everything
then depended on the success of the recall and reissue system. In most
African countries, pupils did not have adequate access to textbooks, in
science as well as other subjects. In Kenya, for example, textbooks were
produced locally. Since there were many books when all the subjects were
included, a significant proportion of students could not afford to buy
them. Teachers ended up writing the content of lessons on the board and
students spent inordinate amounts of time copying material. Where
appropriate written materials were not available, the quality of teaching
and learning was likely to be adversely affected.

These problems extended beyond low-income African countries.
According to recent research in Mexico, students in rural and marginal
urban areas often did not have textbooks. The same appeared to be true
in Chile, where teachers had been known to photocopy their notes, and
sell them at a low price to the pupils or to the pupil/teacher association.
The material available to most disadvantaged pupils was therefore very
limited, and it might have been of dubious quality. In Morocco,
textbooks were produced by the Ministry but distributed commercially.
It was found in the research that a fairly high proportion of the students
had the necessary science textbook (77 per cent of the students at lower
secondary and 60 per cent at upper secondary). This high average
nevertheless hid large disparities by school grade and subjects. A quarter
of the teachers said that half of their pupils did not have a textbook. It is
worth stressing that when only two textbooks are needed for science, it
helps in making them accessible to pupils (Caillods et al., 1997).

Making books available did not necessarily mean that teachers used
them. In Morocco, teachers were fairly critical of the natural sciences
textbooks, and half of them said that they did not use them. They used
other books, particularly foreign ones, instead. Similar results had been
found in Chile by Schiefelbein, Farrell, and Sepuvelda-Stuardo (1983).
They found that science and mathematics teachers were less likely to use

textbooks than any other subject teachers; they also showed that textbooks were least used by teachers with little experience. This lack of enthusiasm could be due to a lack of emphasis on the use of textbooks in the teacher-training programme. The authors concluded that teachers should be trained in how to use them. It might also be that the books themselves were not always of high quality. Both in Morocco and Malaysia, teachers and students were frequently found to use examination practice books ('crammers') to prepare for examinations, instead of officially approved materials.

Little information was available on the quality of the textbooks. Some of them were clearly out-of-date. In Mexico, for example, available textbooks in physics and chemistry were written in the 1950s and 1960s, according to the authors of the IIEP monograph, and thus did not reflect recent developments in pedagogy or science and technology. In Chile, there were several textbooks and their quality varied. Publishers competed to win over teachers and it seemed that teachers might be more sensitive to the cost and publicity than to the quality of the material offered. Other countries systematically upgraded science textbooks and made repeated attempts to improve quality and relevance. This seemed most common where there were national curriculum development facilities which orchestrated the process, as in Malaysia and Korea.

In-service training

Many of the countries studied had introduced curriculum or examination reform in the last five years. Information on curriculum change and staff preparation for its implementation had become major functions of in-service training courses for science teachers. Some countries (e.g. Thailand and to a more limited extent Kenya) had used a 'cascade approach' – i.e. re-training of a certain number of teachers who were charged with training their colleagues in the field – for this purpose. In other countries, e.g. Chile, a large number and variety of institutions – universities, teacher-training institutes, municipalities, teacher associations and even commercial undertakings – had been mobilized to offer residential or distance courses for in-service teacher training accompanying the introduction of a new curriculum.

The case studies on Malaysia and Morocco indicate that large proportions of secondary science teachers in these countries had received some form of in-service training (60 per cent in Malaysia; 85 per cent of science teachers at lower secondary and 30 per cent of those teaching at upper secondary level in Morocco). In both countries, most of these courses were of short duration – two to three days on average. Depending in particular on its specific purpose, teachers' level of satisfaction with the in-service training provided varied: in Malaysia, 87 per cent of the science teachers found the courses accompanying the introduction of the new curriculum useful in terms of learning about new teaching methods. Yet, observation of classroom practice in case-study schools was largely unable to discern many activities that reflected those approaches to the teaching of science that in-service courses were promoting. Whatever teachers were learning in the in-service courses was not clearly and systematically translated into practice. Common criticisms of in-service courses that emerged from the interviews included comments that they were:

- too short to be effective;
- having little effect on teachers' actual practice, because they did not take into account the realities of many schools and students;
- ephemeral, with no follow-up or support materials;
- sometimes run by those without adequate experience themselves;
- concentrated on description rather than the acquisition of new skills (Sharifah Maimunah and Lewin, 1993).

In Morocco, science teachers also indicated that they appreciated the short in-service training sessions which had been organized to the extent that they informed them of new changes in the curricula and showed them new teaching methods. A high proportion of teachers, however, felt that they had not been well trained in methods of pupil evaluation. This was worrying in a system where selection decisions were based wholly or partly on school-based evaluation.

In Morocco, the Ministry furthermore provided teachers with support in the form of 'instructions' – i.e. detailed guidelines specifying the aims and objectives of each course, the content to be taught, and the associated methods of evaluation. A high proportion of secondary science teachers had such 'official instructions'. In the IIEP survey it appeared that lower secondary school teachers found them useful but upper secondary school teachers were more critical (Caillods, Göttelmann-Duret et al., 1997).

Officially, all other countries, including those which had not introduced curriculum reforms recently, claimed that they organized in-service training but no detailed information on such programmes was easily available. Nor was any available on the cost-effectiveness of investment in in-service activity – a subject returned to in *Chapter IV*.

4. Specialized facilities and equipment

In all the countries studied, secondary schools were supposed to have at least one multi-purpose laboratory at the lower secondary school level and one or more laboratories per subject, according to the school size, at the upper secondary school level. The reality in the field was quite different. Eighty per cent of secondary schools in Burkina Faso, 50 per cent of schools in Kenya and 20 per cent of middle schools in Senegal had no laboratory at all. In Chile, 50 per cent of the schools had no laboratory and those that existed were, in the majority of cases, poorly equipped. In Korea, 72 per cent of middle schools and 55 per cent of senior secondary schools did not have enough laboratories; lack of equipment was also said to prevent teaching with an emphasis on practical activities.

Numerous school-mapping exercises conducted by IIEP in over 10 countries of Africa and Latin America have shown that a large number of lower secondary schools in rural areas, particularly those which had been built recently without the support of an aid agency, did not have laboratory or specialized science facilities; equipment and chemical reagents were everywhere in short supply; running water, in some cases, was not available in specialized rooms. Similar observations have been made in numerous sector studies.

Malaysia and Morocco were more fortunate than many other countries in this respect. Their schools were well equipped with laboratories and specialized classrooms, even if there were some differences between regions. But this did not necessarily mean that the facilities were well used. In Morocco, teachers complained that they did not have enough equipment but this was partly due to the lack of flexibility in the syllabus – all the classes needed the same equipment more or less at the same time. Given the very high cost of laboratories, a number of countries used kits in addition to or instead of having laboratories. Science kits are discussed in more detail in *Chapter IV*.

Even where there were a sufficient number of laboratories and enough equipment, *maintenance* was a major problem. *Laboratory assistants* and attendants are usually responsible for minor repairs, in addition to preparing equipment, chemicals and biological material for laboratory lessons. Few countries had laboratory assistants (see *Table 2.3*). When qualified teachers were free to do so, a significant proportion of their time was spent on low-level tasks which might otherwise have been available for additional science teaching. Most teachers, however, did not have the time nor had they been trained to carry out maintenance. Where laboratory assistants were provided it is not altogether clear what determined their deployment. Moreover, they had often only been trained on-the-job. Those who had two or three years of post-secondary science but had a lower status than science teachers felt inevitably rather frustrated. In the Moroccan case study it was not possible to conclude that there were fewer problems of maintenance in those schools which had laboratory assistants than in others as the latters' duties were not always clearly defined (Caillods, Göttelmann-Duret et al.,1997).

To deal with the problem of maintenance, the Moroccan Ministry of Education created a National Centre for Educational Technologies. This centre, responsible for the distribution and maintenance of equipment, was also expected to produce local equipment. As a whole, however, the maintenance, repair and replacement of broken pieces of equipment was a critical issue everywhere.

Section 4: Science teaching in practice

Numerous studies conducted in developed as well as in developing countries stress the fact that science teaching is seldom linked to the developing of thinking skills related to solving real problems. It is more often based on rote learning and theoretical exercises. Observation and experimentation does occur but often it is not clear what the purposes are or what outcomes are valued. All the monographs written in the framework of the research programme contain indications of similar criticisms. Although every country had schools where science appeared to be taught very effectively, they were in the minority. In practice, much teaching fell short of curricula expectations, failed to develop intellectual skills related to science systematically, and where practical activity is possible often overvalued this, at the expense of the reasoning skills involved in experimentation. Lack of facilities and equipment and inappropriateness of the teaching and learning conditions were advanced as major reasons for the teaching methods used in practice, but this may often conceal problems which are grounded in poor quality teacher preparation and low motivation to teach science as effectively as possible with limited resources.

Class size

Class size was in many cases considered as a major stumbling block to the practising of regular group activity and pupil-centered science teaching. The number of pupils per class actually varied a great deal from country to country (*Table 2.3*). It was very high in most African countries

(Burkina Faso, Senegal and Kenya) – on average 64 pupils per class in state lower secondary schools in Burkina Faso and 60 in private ones, and 55 in the experimental science stream at upper secondary level.

Most research on class size fails to identify learning achievement differences related to the number of pupils in the class. Often, the schools with the smallest classes have the poorest results, because they tend to be remote rural schools or are under-enrolled urban schools for which there may be weak demand because of their bad results. Having a small number of pupils in a class does not necessarily mean that the teaching is of high quality; conversely, a large number does not necessarily mean that teaching is ineffective. Korea, which had very large average class sizes (50 to 54 students per class), managed to outperform other countries in the IEA science tests. This may be because the effects of class size are masked by many other intervening factors which cannot easily be controlled. It may also be because measures of achievement are insensitive to outcomes which may have benefited from smaller class size (e.g. ability to undertake small group experiments). Assuming that most examinations do measure some useful outcomes, and noting that examination performance is not sensitive to class size within a wide range, the conclusion for planners is that class size can be maximized – consistent with physical limits of space – without affecting the measured outcome of science education.

Several conditions have to be met, however, for successful teaching of large classes – suitably trained teachers, sufficient space, books and equipment, motivated students and appropriate curriculum design, etc. – and these are not satisfied in many poorer developing countries. Furthermore, it is questionable whether some kinds of practical work can be undertaken with large classes (more than, for example, 40). Science classes are sometimes subdivided to create smaller groups for laboratory work. This was the case in Morocco, Senegal, Papua New Guinea and France. Elsewhere, science was not considered as a workshop subject, like design and technology classes, which often had class size limitation. Student activity groups were likely to be large (although, often experiments are designed for small groups), equipment demands would be extensive, and classroom management might become very difficult. In larger classes it is probably difficult for many students to even see the demonstrations, depending on how they are presented.

Actual teaching practice

The case studies carried out in Morocco and Malaysia attempted to investigate actual science teaching practices and identify the factors which tended to impede effective practical activity. In Morocco, teachers were asked to describe how many periods for a particular class (the one which they taught most) they spent in a specialized classroom, how many experiments (demonstrations) they organized themselves during the month of January 1993, and how many sessions they organized of practical work for students. The findings were quite interesting. Although a high proportion of teachers held all of their lessons in a specialized classroom, they did not do many demonstrations – on average six at lower secondary level, and on average three at upper secondary level per month. It was in the first year of lower secondary (Grade 9) that teachers did the most demonstrations on average (seven in the month) and in the last year of upper secondary, in the so-called experimental stream, that they did the fewest! Only 3 per cent of the teachers of lower secondary did no experiments during that month, but between a quarter and a third of the teachers of the experimental science stream in the last two years of upper secondary had not done any. It seemed that the older the pupils and the more advanced in their studies (particularly those who specialized in science), the more they were expected to learn science through memorization, theoretical exposition and reasoning (Caillods et al., 1997).

In terms of students doing practical work in groups or individually, the situation was just as worrying – more than one-third of the teachers did not organize any at lower secondary level and three-quarters of the teachers did not do so at the upper secondary level. A good number of teachers did not split the class into two groups although the facilities existed, probably because they did not organize practical work. The reasons they gave to explain their teaching methods included lack of equipment, content overload of the curriculum and pressure from the examinations (which, in Morocco, ran through the last three years of secondary education). According to teachers, there was so much to be covered in the programme that three-quarters of those who taught at upper secondary level felt it necessary to organize extra lessons for their students free of charge. The laboratories were there, a good deal of equipment was there, but the content of the curriculum and the organization of examinations was said to be such that teaching had to focus on cramming for the examinations. Very few teachers organized science projects at the school level which might have helped to interrelate the different science disciplines and could have helped in motivating students. More frequently they participated in the national science 'Olympics' competition (70 per cent at the upper secondary level).

It was not possible to test whether practical activity was more common in high-performing schools in Morocco. Pass rates were high everywhere and tests were not comparable from one region to another. There was a weak indication that the schools with the highest success rates were not necessarily those with the best facilities, nor those where teachers organized the largest number of practical sessions. There could be many reasons for this, including the fact that sessions of practical work might not be thought to provide meaningful learning.

The Malaysian study suggested that most teachers had knowledge-based views of science education and its purposes and this was reflected in their teaching methods. Lessons observed were generally didactically presented and were teacher-centred with little input from students. Typically, lessons started with a discussion of previous work, followed by instructions for a practical activity or a demonstration which was then written up. Students in most cases played a passive role – listening, following instructions, or copying from the blackboard. Although some teachers used question and answer techniques in class, questions tended to be confined to lower cognitive levels, mostly involving the recall of information rather than requiring reasoning and interpretation. There seemed to be real problems with the guided discovery-inquiry approach advocated by the curriculum both at lower and upper secondary levels. Observations of teaching provided few examples of what seemed to be guided discovery lessons.

Fortunately, there were some exceptions. These were mainly found in the residential schools, where it was more frequent for students to be asked questions and to take part in discussing how to do experiments and interpreting the results. Teachers who did this noted that it was more time-consuming than 'chalk and talk' and that a lot of effort had to be invested in preparing lessons and thinking about questioning strategies to lead students through a scientific reasoning process. The 'conveying information' approach emphasized by many teachers was consistent with views held on public examinations. Many teachers chose to stress factual knowledge, since they felt this was what the examinations predominantly tested. As one teacher said, "Even if students do not understand concepts, they can get through the examination by merely memorizing facts".

The following excerpts illustrate some of the problems commonly attached to practical work, even in the best circumstances when teachers have the necessary facilities to organize practical work and when they actually do it. Teachers favour group activity rather than individual activity because it is easier to maintain discipline and to control the groups. However, the groups are often too big to allow an even spread of individual learning experiences.

Box 2.1: Group practical activity in science in a Malaysian school

In a practical lesson conducted in School 2, students were required to test the pH of several substances. The experiment required very simple apparatus consisting only of test tubes and pH paper, yet it was done in groups of five, with only two students involved in the testing of each substance. Although the teacher did attempt to ensure that each substance was tested by a different pair of students, this did not seem to work and ultimately the same students were involved again and again. Other students were observed to be occupying themselves by copying what was on the blackboard, chatting among themselves or merely passively observing the experiment.

Source: Sharifah Maimunah and Lewin, 1993 (120).

Another problem was concerned with the use that teachers made of the observations and whether or not it became a real learning experience.

Box 2.2: Discussion of the aims of practical activity in a Malaysian school

In School 7, an experiment was conducted in which students were required to investigate how water enabled acids and alkalis to exhibit their specific properties. To do this, students were required to do a litmus paper test on the aqueous and non-aqueous forms of benzoic acid as well as on benzoic acid dissolved in toluene. The same steps were to be repeated with sodium hydroxide. At the beginning of the experiment, the teacher merely mentioned that the experiment was aimed at investigating the role played by water. Despite the fact that the experiment was conducted in several parts, no attempt was made to link the various parts in the discussion of the aims. As a result, students were observed to be going about the practical activity mechanically without understanding the relationship between the various parts of the experiment. Although teachers were able to help students acquire manipulative skills in handling apparatus, making observations, and recording data, they seemed less able to encourage students to acquire higher order intellectual skills, e.g. interpreting data, hypothesizing, making generalizations. This was reflected in the infrequency of meaningful discussions based around practical activity. More often than not, discussions were merely a comparison of observations and experimental results between groups. The teacher then generally read out to the students the anticipated conclusion of the experiment, which students copied into their books. There was rarely a discussion of why results might differ between groups. When the experiments failed, many teachers would simply suggest to students that they observe another group to see what happened when the experiment worked well, without explaining why it might have failed in that instance.

Source: Sharifah Maimunah and Lewin, 1993 (121).

The two examples from Malaysia given above illustrate a variety of teaching practices in the best physical and financial cases. In many cases, science teaching was undertaken by teachers who had not themselves benefited from effective teaching.

How teachers teach is a function of how adequate their knowledge of science is and the sophistication of their conceptions of science constructs; it also depends on their interpersonal skills with students, their pedagogical skill, the leadership they receive and the learning culture in schools. All these factors play a part in determining how teachers actually teach. It may take more than one generation of science teachers, and supportive changes in other areas, before some of the expectations of curriculum developers for science teaching are met.

This does not preclude the notion that some teachers know how to transform an experiment into a real learning experience. But many of these teachers are to be found in selective schools which combine the best students with the best facilities. These included the residential schools in Malaysia, the best private schools in Chile, Argentina and Brazil, the private schools and the ones run by universities in Mexico, the fully funded government schools in Kenya and the national high schools in Papua New Guinea. Every country had élite schools which differed markedly from typical rural schools. It was observed that these differences were often more marked where decentralized systems of administration and supervision existed.

Section 5:   Cost of science education

Science education is one of the most expensive disciplines to teach in secondary schools, crafts and vocational technical subjects apart.

First, teaching costs are generally higher than those for other subjects. Science teachers are more expensive to train than other teachers as special facilities are required at the higher education level to train them; their cadre is supplemented in some countries where they are in short supply with expensive expatriates. Teaching groups in science are sometimes smaller than for other subjects (as a result of limited laboratory space, or of the relative unpopularity of certain options or streams); the number of teaching periods of science teachers may also, in some countries, be lower than for other subject teachers (as a result of the time they invest in preparation and follow up of practical activity), and teacher utilization rates are therefore lower, leading to greater costs per student.

Second, the recurrent cost burden of science teaching is likely to be substantial. The consumption of analytic grade chemicals is expensive; the provision of consumables for all the sciences often involves foreign exchange costs; and repair, maintenance and replacement costs of equipment are substantial. In addition, laboratory assistants and attendants are appointed in some countries – although, as was seen, not everywhere and not in sufficient number – to undertake laboratory housekeeping, lesson preparation and cleaning up.

Third, science learning materials may be more expensive than other textbooks, especially if they have to be imported.

Fourth, the costs of science education may be unnecessarily high where management and support systems for science education are inadequate or non-existent – the quality of teaching being sensitive to effective management and supervision (e.g. where many teachers are underqualified).

Finally, practical facilities are expensive to build and equip.

In the IIEP project no attempt was made to compute the overall cost of science secondary education. Such information was difficult to obtain without making a detailed analysis of each school budget. At central level, science teachers' salaries were included with those of other secondary teachers; also, non-salary expenditures and capital expenditures frequently appeared in the same accounting line as technical and vocational education. Some information was collected, however, on the potential additional costs of science education by reviewing data on the costs of foreign teachers, construction and equipment of laboratories, and science school textbooks.

1. Science teacher costs

The information collected in the course of the research programme supported the fact that salary costs were higher in science than in other subjects.

In Botswana, Burkina Faso, Senegal, Gabon, and Papua New Guinea, unit costs of science teaching were inflated by the high costs of expatriate salaries. In Papua New Guinea, for example, foreign teachers, who received in principle a salary identical to that of national teachers, sometimes ended up costing five times as much, once their travel costs and that of their family plus an international market allowance had been taken into consideration. The nature of the problem varied, however, depending on the extent to which costs were met by domestic sources or foreign assistance. Even where salaries fell on aid programmes, local costs (housing, transport, allowances) sometimes exceeded the cost of a national teacher. Where the country was paying the whole cost and the proportion of expatriate teachers was high, the incentive to replace such teachers with cheaper expatriates, as Botswana did, and later on with nationals, was very high.

As far as national science teachers were concerned, in most countries they were employed on the same basis as other teaching staff. There were only a few cases where qualified science teachers started several increments higher on the salary scale than other teachers (for example, Botswana). Some countries, such as Morocco, used to give an additional allowance to science teachers in order to attract them to the profession, but this had not turned out to be a sufficient incentive. Overproducing science graduates actually proved to be the most efficient method of securing teachers at sustainable costs.

Splitting classes into groups for laboratory work, as mentioned above, was common in a very limited number of countries. The reduced number of pupils per class occurred more frequently by splitting the groups into different sub-groups for the teaching of different options, particularly when those options were not very popular (e.g. physics, earth science). Where these groups were very small, unit costs were high. Additional recurrent costs arose as a result of the employment of laboratory assistants where these were provided.

Over and above this, came three main additional costs – the maintenance costs of laboratories, the replacement of small pieces of equipment (glassware, for example), and the provision of consumable materials which were used during practical activities. These costs could be substantial and often involved a significant foreign exchange component. They were likely to depend on the number of students enrolled at different levels and in different streams.

There was also almost certainly a large variation in the costs of consumable material between countries, but no systematic data were available on this. In Malaysia, all secondary schools received a per capita grant to cover non-salary expenses involved in the teaching of science and mathematics. The grant, which was computed according to the number of pupils in each grade, ranged from $15 per student in lower secondary, $30 per student in upper secondary and $45 per student in form 6[5]. Not all countries however had such a system, nor were they always able to implement it.

---

5. The grant failed to take into account the number of pupils enrolled in each stream, and this has led to under-estimating the resources going to schools with a high proportion of science students.

Overall it is reasonable to assume that the recurrent costs of teaching science were about 10 to 25 per cent above other academic subjects.

2. The cost of textbooks

The cost of science textbooks varied from US$1 in Korea to US$32 for certain books in Senegal and Burkina Faso. Whenever a book was produced outside the country and imported, its price was very high – beyond the reach of pupils, of teachers and sometimes even of schools. It was in the very poorest countries, where there was little local infrastructure and where local markets were limited, that the science textbooks were the most expensive. The variation in the price of books could also be explained by the different quality of the paper (the price of which had risen rapidly in past years, affecting the price of books even in those countries which produced their own), and the type of reproduction (of photographs, diagrams, etc.). There was no evidence that science books were indeed more expensive than those for other subjects, except that they were more often imported.

Taking into account the fact that a pupil was expected to buy not one but several science textbooks (one per subject, plus students' exercise books, worksheets, practical books, reference books, etc.), the cost of science textbooks ended up very high per pupil, ranging from $3.7 in lower secondary in Morocco, $10.6 in Thai lower secondary, $14.5 in Thai upper secondary, to $28 and $77 for the lower secondary school pupils and upper secondary school pupils of Burkina Faso (Caillods, Göttelmann-Duret, 1991). This was just for one subject – science – leaving aside all the others: it was not surprising to discover that pupils did not have the required textbooks.

3. The costs of laboratories

The research data collected indicated an extraordinarily wide range in laboratory costs between and within countries. This variation reflected very different expectations of the standards to which facilities should be provided, and the different costs of building, materials and equipment in different locations.

Table 2.4 gives some idea of how laboratory costs vary in different countries for a fully equipped physics laboratory at upper secondary level with a preparation room facility.

Table 2.4   Estimated costs of a fully equipped physics laboratory with a preparation room (1990 US$)

Country	Laboratory cost	Equipment	Total
Argentina	35 000	10 000	45 000
Botswana Upper sec.	132 000	12 120	144 120
Junior sec.	68 700	7 900	76 600
Burkina Faso	21 400	6 700	28 100
Chile	13 000	6 250	19 250
Jordan	16 000	10 300	26 300
Kenya local	32 000	9 500	41 500
external aid	64 000	9 500	73 500
Korea Academic HS	72 000	21 900	93 900
Science schools	72 000	62 400	134 400
Morocco	29 000	46 900	75 900
Papua N.G. Upper sec	95 000	20 000	115 000
Junior sec.	76 500	18 000	94 500
Senegal	56 500	14 000	70 500
Thailand ordinary	24 500	3 500	28 000
community	18 642	3 500	22 142
Malaysia	24 000	7 400	31 400
Mexico	27 886	nd	nd

nd = no data

Note: the costs have been 'standardized' for the sake of this presentation to correct for the fact that, in some countries, laboratories, particularly if they were built with the help of an aid agency, were part of a Scientific Block or Unit. *Source* : Caillods, Göttelmann-Duret, 1991.

The figures above indicate large variations between countries in the type of specialized classroom constructed, the type of building, fitting, equipment, and furniture included. The lowest cost ratios were found where science rooms – multipurpose specialized rooms rather than science laboratories – were being constructed. In these rooms, basic services were usually available (electricity, water, etc.), and the equipment provided was simple and relatively inexpensive without specialized high-cost items.   This was the case in Chile, where only multipurpose specialized classrooms were built, without even a preparation room. A corner in the room was simply identified where the teacher could prepare what he/she needed. The highest cost in the list of countries appears to be in Botswana, where a senior secondary school laboratory included suitable benches for laboratory work, enough sockets, gas outlets, water taps and sinks for 10 to 12 groups, a built-in variable low-voltage power supply for the teacher and 12 work stations. Recently, the Ministry of Education was looking into the possibility of reducing the costs of such laboratories and the equipment, particularly at the lower secondary level.

The ratio of the cost of a laboratory to that of a normal classroom appeared to cover a range from about 1.6 to 5 (or even more than 8 if one considered the cost of a full scientific unit) (Caillods, Göttelmann-Duret, 1991). Where this ratio was highest it was tempting to conclude that science facilities were being over-engineered in relation to typical facilities for normal subject teaching.

The figures given above, in fact, conceal large variations according to the financing source and the location within the same country. Remote rural areas may have building costs two or three times those in accessible urban locations for the same building, due to the transportation costs. In some cases, rural costs may be much cheaper if built to a much lower standard. The cost of land may have a strong influence on total costs; it is not included here. This is usually not a factor in rural areas.

Some figures from Kenya illustrate how costs can vary in a country. The cost in 1990 US$ for a high-cost donor-financed laboratory and a lower-cost building constructed to normal Ministry standards was as shown in *Table 2.5*.

Table 2.5   Cost of a science laboratory in Kenya (US$ 1990)

Location	High cost	Low cost
Near Nairobi	32 000	20 000
Rural (near tarmac road)	64 000	32 000
Rural (off tarmac road)	80 000	40 000

Source: Obura, 1992.

These estimates are much higher than those given for the construction of a specialized classroom built by the Parent-Teacher Association of a rural government-aided school, basically an ordinary classroom with minimum interior fittings and electricity used only as a service. In 1990, about $2,000 was the cost of a typical, ordinary classroom in Kenya.

The equipment imported for high-cost donor-supported laboratories cost about $9,500 per laboratory. At the other end of the spectrum, locally produced science kits from the Science Equipment Project Unit cost about $600 to cover all three science subjects. These kits were designed to be used in schools without special laboratory facilities.

Considering the common pattern of use which was discussed above, as well as the limited capacity of some of the countries to support their running costs, these high laboratory building and running costs raise a number of questions: were they justified by pedagogical necessity, and demonstrated learning gains? Could they not be provided at lower cost? These issues are discussed in more detail in *Chapter 4*, along with the attractions of utilizing more low-cost methods for delivering practically based learning experiences.

Section 6: How much science do secondary pupils know?

Science education is expensive and regarded as strategically important to development. Policy-makers, parents, teachers and students then all have an interest in whether investments in science education are effective. The most obvious indicator of this is an assessment of the science knowledge and skills that pupils actually have on completing various stages of their formal schooling. Yet most countries lack appropriate information for the monitoring and possible improvement of science achievement. Standardized examination results constitute an insufficient indicator. Information on raw scores would be needed. The most recent data available concern those countries which participated in the IEA (International Association for the Evaluation of Educational Achievement) and the IAEP (The International Assessment of Educational Progress) science studies. Although, what is of particular significance for planners and decision-makers would be the analysis at the intra-national level, it is relevant to discuss here some of the results of international studies – including those which try to unravel 'determinants of science achievement'.

1.   Present difficulties in assessing actual science achievements

Pupils' examination pass rates and average scores in the different science subjects are generally the only 'outcome' data which are easily available at national level. The existing data concern the Primary Leaving Examination, on the one hand, and the major examination at the end of secondary education (Secondary Leaving or University Entrance Examination), on the other. In most countries, there is little or no information allowing for nationwide assessment of pupils' science achievements at intermediate levels. (Among the countries included in the IIEP's project, only Botswana, Burkina Faso, Papua New Guinea,

Senegal and Malaysia[6] organize national examinations at the end of lower secondary). This may be more serious where the end of lower secondary constitutes the termination of basic education and the actual school-leaving level for a majority of children.

Standardized secondary leaving (or University Entrance) examinations exist in many countries. Examination pass rates or standardized scores in science subjects that students obtain at this level may, however, constitute poor performance indicators; pass rates depend on the number of study places that are available (in these subjects) at a subsequent level of education. They may also reflect the existing policy on controlling and shaping student flows. Hence, the meaning of a pass standard may change from year to year. The analysis of science *baccalauréat* pass rates and average scores over a period of time by region/*Académie* and by subject in Morocco revealed, for example, such inconsistencies and fluctuations that no conclusion on the evolution of science achievements of secondary school leavers could be drawn (Caillods, Göttelmann-Duret, 1997). Actual achievement levels would be better assessed through the analysis of students' raw scores or of the scores obtained in national tests administered for purely evaluative purposes. From the data available and those published in other studies, some observations on levels of achievement in science can be advanced.

2. Insights into levels and patterns of science achievements

Assessment tests in science subjects covering several developed and developing countries have been conducted by two specialized organizations, namely the IEA and the IAEP educational testing service. Many bilateral aid agencies have also begun to carry out assessment of pupil achievement (in science and other areas) in various developing countries.

Both specialized international associations mentioned above have paid particular attention to science achievements amongst nine-year-olds and the 13 or 14-year age group. In many countries, these are the ages at which pupils are expected to reach the end of elementary and lower secondary education, respectively. The IEA also tested science achievements of students who were completing the secondary cycle and who had specialized in science at this level. Some of the main results and conclusions from these international studies regarding the two latter groups are summarized in *Table 2.6*.

6. This has in 1995 become a monitoring assessment rather than a selection examination.

Table 2.6   Classification of countries by score group for total and sub-
scores (13-14 years old)

Score range	Biology (23 items)	Chemistry (15 items)	Physics (23 items)
More than 65%	HUN JAP	HUN	CHN HUN JAP NET
60-65%	CAE CAF FIN IT9 NET POL THA	CHN NET	AUS CAE CAF FIN IT9 KOR NOR SW8
55-59%	AUS ISR KOR NOR PNG SIN SW8	JAP FIN ISR SW8 POL NOR	ENG HKO ISR PNG POL SIN SW7 THA
50-54%	CHN ENG HKO IT8 SW7	CAE KOR SW7 IT9 AUS	IT8
Less than 50%	GHA NIG PHI ZIM	CAF ENG GHA HKO IT8 NIG PHI PNG SIN THA ZIM	GHA NIG PHI ZIM
	Median 58.5%	Median 50.5%	Median 59.5%

AUS - Australia; CAE - Canada (English-speaking); CAF - Canada (French speaking); CHN - China; ENG -England; FIN - Finland; GHA - Ghana; HKO - Hong Kong; HUN - Hungary; ISR - Israël; IT8/IT9 - Italy; JAP - Japan; KOR - Korea; NET - Netherlands; NIG - Nigeria; NOR - Norway; PNG - Papua New Guinea; PHI - Philippines; POL - Poland; SIN - Singapore; SW7/SW8 - Sweden; THA - Thailand; ZIM - Zimbabwe.

Source: Postlethwaite and Wiley, 1992.

2.1   *Science achievements of the 13/14-year-olds*

The average score in the probability sample of the 13/14-year age group, obtained in the core science tests administered by the IEA, varied considerably from one country to another; the range was from less than

40 per cent to slightly more than 70 per cent of correct answers. It is worth noting that in a good number of countries fewer than 50 per cent of correct answers were received. At the same time, however, the ranking of countries varied when their average sub-scores in the different science subjects were considered. Japan, for example, had high total achievement scores but slightly lower scores in chemistry than in biology and physics; Thailand scored well in biology but poorly in chemistry. In some countries, namely Hungary, the 14-year-olds scored consistently very well. The results of both the IEA and the IAEP science tests among 13-year-olds, furthermore, indicated relatively high average scores for Korea (Japan was not included in the IAEP test) (see Postlethwaite and Wiley, 1992; Lapointe, Mead and Askew, 1992b).

Policy-makers and educators may learn from international comparisons of total performance levels in science about what can be achieved in certain contexts. However, what they need to know most of all are the characteristics of low, high and middle-level science achievers in their *own* country and what could be done to improve their performance. Indeed, science achievement data must be interpreted within the specific context of each country. Some education systems are less well developed than others (when the IEA science tests were conducted, the percentage of the 14-year age group in school ranged from 11 per cent in Papua New Guinea to more than 90 per cent in Korea and Japan); different options as regards the organization and content of the curriculum may have different effects on achievement; a wide range of school-related and other factors – which vary from one country to another – influence science learning. Interesting insights could, therefore, be gained from a more differentiated analysis of achievement data, which would look separately, for example, at achievement patterns of different student populations (boys and girls, high and low achievers, etc.), and explore the impact of different variables on their achievement in science.

It was interesting to note that there was evidence of high achievement in science amongst 13/14 year-olds in almost all countries that participated in the IEA and/or the IAEP science tests, as demonstrated by the performances of the 'top 10 per cent' of the pupil populations investigated in the latter, and those of the top 20 per cent recorded in the former. At the other end of the scale, low achievement levels in science could also be observed everywhere. The science performance level of the 10 to 20 per cent of the student population who were the lowest scorers varied a great deal among the countries: Korean pupils in the lowest decile performed at about the same level as the average pupils in Jordan. In some countries, the science achievement levels of the bottom 10 or 20 per cent were particularly alarming; the IEA test revealed that the bottom

20 per cent of 14-year-olds investigated in Ghana, Nigeria, the Philippines and Zimbabwe scored at a level just above chance (i.e. they could be as successful by ticking at random the possible answers to the multiple choice test items given). In some other countries in which relatively high *average* achievement scores in science were recorded – e.g. Hong Kong, Singapore and Sweden – the bottom group showed very low achievement as well. As the IEA study pointed out, the low scientific literacy of these bottom groups poses a problem for educational planners in the countries concerned since these low achievers are most likely to enter the labour market directly without any additional basic science education.

Another significant result of the studies cited was that in nearly all countries covered, 13 or 14-year-old boys performed better in the science tests administered than girls of the same age. There were some exceptions, however, showing that girls did not necessarily perform less well than boys in science.

A further conclusion to which educational planners may want to give particular attention is the fact that national average scores can hide large achievement disparities within a country, according to the school or the class which the pupil attends. The IEA study found that in Ghana and the Philippines, as well as in Italy and the Netherlands, the science scores of the 14-year-olds correlated highly with the school or class attended, while the Nordic countries and Japan recorded much more homogeneous science achievements across classes and schools. Societal factors, organization of the school system and other variables may explain these variations. Clearly, it is important to appreciate these context variables when considering policy interventions.

In any case, very low science performance levels amongst the bottom 20 per cent achievers as well as significant achievement disparities between boys and girls, and pupils from different schools, provide warning signals of a problem. They are likely to be at variance with aims of equitable access, the attainment of some level of scientific literacy for all, and also suggest wasted investment.

2.2 *Achievements of secondary school leavers specialized in science*

The difficulties confronting international comparisons for those students completing secondary education with some specialization in science subjects were even greater than the ones already noted for the 13/14-year-old age group. These students constituted a very restricted and selected group. The proportion of the age group studying science (as a percentage of the age group in school) at this level varied between about 1 per cent in Burkina Faso and Papua New Guinea, to almost 25 per cent in Korea, according to the IIEP survey (Caillods, Göttelmann-Duret, 1991). A central question was to what extent the participation rate in specialized secondary science studies affected the achievement levels of those enrolled in those programmes and/or the performance levels of the very high achievers. Other critical questions were whether the performance levels differed according to the grade level considered (in some countries, secondary completers had gone through 12 years, in others, through 13 years of schooling) and according to the number of science subjects that specialized students studied at that level.

With respect to these issues, the results for Ghana found in the IEA science achievement study are particularly worthy of note: performance levels for Ghananian secondary school completers specialized in science were quite high, while the achievement levels found for the 14-year-old group as a whole were very low. This phenomenon may be explained by two factors which have a general impact on science achievements at this level, namely the percentage of the age group enrolled in science (which ranged in Ghana from between 0.2 per cent and 1 per cent, depending on the science subject considered) and the grade level attended by the students tested (some of the Ghanaian students tested were attending Grade 13, while those in the other countries mostly attended Grade 12).

The IEA study concluded that the higher the proportion of the age group enrolled in biology, chemistry and physics, the lower the average score recorded in these subjects (see Postlethwaite and Wiley, 1992). However, it found no evidence of a relationship between the percentage of an age group studying a science subject and the performance of the élite in this subject, in other words, the science achievement of the 'top 10 per cent' of senior secondary students specializing in science can be high irrespective of the proportion of the age-group enrolling in these subjects. The study also suggested that there was a tendency for countries where fewer science subjects were studied at upper secondary levels to score higher in these subjects; at the same time, however, it pointed to the case of certain 'high performers', such as Hong Kong, which indicated that increasing the number of subjects studied from three or four to five or six could be achieved without affecting achievement.

3.   Factors affecting science achievement

According to various authors, a certain number of parameters, apart from those already mentioned above, were commonly believed to have a positive effect on achievement in science (most research refers to the 13/14-year or the 9-year age groups) (see, e.g. Walberg, 1991).

Firstly, achievements in science were positively related to the amount of time that pupils engaged in science learning (at school, at home and elsewhere, e.g. in private tuition) in most countries. This conclusion could be drawn from the IAEP international science study (see *Table 2.7*) and from a review of 60 World Bank studies on factors having positive effects on learning outcomes in science and in other subject matters (Walberg, 1991). The IEA study also found a positive effect regarding the amount of homework (see *Table 2.7*).

Furthermore, the number and appropriateness of curriculum topics to which students were exposed had also proved to have powerful effects on achievement. The first International Science Study carried out by the IEA in the early 1970s had already noted these effects of exposure to learning; the same conclusion appears in the 1992 IAEP mathematics test for 17-year-olds. This result is not surprising. It is more difficult to decide at what level diminishing returns set in and additional time allocation produces little achievement gain.

Studies on more or less significant factors of school teaching and learning in a variety of subjects had also pointed to the positive impact of the quality and availability of textbooks and other teaching and learning materials; there was also some evidence that the level of subject-matter knowledge and subject matter-related pedagogical skills of teachers affected the quality of instruction and its outcome (see Walberg, 1991). However, there are presently no consistent research results on the exact impact of some of these factors on science achievement. The same holds true for certain complex variables such as the 'quality of school functioning' or 'school climate' which are difficult to determine but seem to have a significant effect on pupil achievement – in science as in other subjects. There appears to be a positive relation between students' views on science studies and their achievements in science subjects. *Table 2.8* indicates the strength of this relationship in all countries participating in the IEA study.

Surprisingly, certain parameters, such as class size and science laboratory work, appeared significant in only a few countries. According to both the IEA and the IAEP studies, in some countries there was no statistically significant positive relationship between the amount of practical experiments that students had conducted in the classrooms and their achievement in science (see *Tables 2.7* and *2.8*).

Table 2.7 Science achievements of 13-year old students: relationship of classroom factors and average scores (average percentage of correct answers within the population)

Comprehensive populations	Amount of listening to science lessons[1]	Amount of student conducted experiments[1]	Amount of science testing[1]	Amount of time spent on science homework[1]
Korea	+	-	-	-
Taiwan	+	0	+	+
Switzerland 14 cantons	0	0	0	-
Hungary	0	-	0	+
Soviet Union Russian-speaking schools in 14 republics	+	0	+	+
Slovenia	+	-	-	0
Emilia-Romagna, Italy †	0	0	-	0
Israel Hebrew-speaking schools	+	-	0	0
Canada	0	+	+	0
France	+	-	0	-
Scotland †	0	0	0	0
Spain Spanish-speaking schools except in Cataluña	0	-	0	+
United States †	0	-	0	0
Ireland	+	0	0	+
Jordan	+	-	0	0
Populations with exclusions or low participation				
England Low participation ‡	-	0	-	+
China in-school population, restricted grades, 20 provinces & cities	+	0	0	0
Portugal in-school population, restricted grades †	0	-	0	0
São Paulo, Brazil Restricted grades	0	-	-	0
Fortaleza, Brazil in-school population, restricted grades	0	-	0	+

+ Statistically significant positive linear relationship.
- Statistically significant negative linear relationship. 0 No statistically signifiant linear relationship.
† Combined school and student participation rate is below 80 but at least 70; interpret results with caution because of possible nonresponse bias.
‡ Combined school and student participation rate is below 70; interpret results with extreme caution because of possible nonresponse bias.   [1] IAEP Student Questionnaire, Age 13.
Source: Lapointe, Mead and Askew, 1992b.

73

Table 2.8   The effects of different variables on science achievement

Country	SES of home	Literacy of home	Sex of student	Like school	Teaching style	Experi-ments	Classroom effort	Home-work	Sch. Sc. Equipment	Views of and interest in science	R²
Australia	+++	+++	--		--	+		+		+	.35
China	++	++	-		-	+	+	++	++	+	.23
England	+++	+++	--		--	+		++	+	++	.37
Finland	++	++	-	+	-	+		--		+++	.19
Hong Kong	+	+	--		--	+++		++	++	++	.28
Hungary	+++	+++	--	+	--					++	.23
Italy (Grade 8)	+++	+++	--			-	+	++	+	+	.31
Italy (Grade 9)	+++	+++	--		+	-	++	++	++	--	.16
Japan	na	+++	--	++	--	+	+++	+		+	.27
Korea	++	+++	--	+	-	-	+		+++	+++	.36
Nether-lands	+	++	---	+	--		++	+	++		.37
Nigeria	++	++	--	+	-	-		++		+++	.15
Norway	na	+++	---	na	--	+++	na		+	na	.16
Philip-pines	+++	++	--	++	--	+	+	+	+	+++	.27
Poland	++	++	--		-			+		++	.12
Singapore	+++	+++	--		-	+		++		+++	.40
Sweden (Grade 7)	++	+++	--	+	-	+++	++			+	.23
Sweden (Grade 8)	++	++	--	+	-	+	++	-		+++	.27
Thailand	+	+	--		---	++	+	++	-	++	.17
Signs for 19 countries	17+	19+	19-	9+	18- 1+	12+ 3-	10+	13+ 2-	8+ 1-	15+ 1-	

Table 2.8 presents the results of the influences of ten constructs on science achievement. Using the conventions of a single + (plus) or - (minus) sign representing a weak effect (+: .05 to .09; -: -.05 to -.09), ++ or - representing an effect (++: .10 to .19; --: -.10 to -.19), and +++ or --- representing a strong effect (+++: .20 or higher; ---: -.20 or lower) the table was constructed to indicate the effects on student science achievement within each country.

(R²) indicates the amount of variance in science achievement among students explained by the variables included. The convention 'na' means that no data were available and no entry means that the coefficient lay between ± 04, indicating that there was no effect. SES= Socio-economic status.

Source: Postlethwaite and Wiley, 1992.

In Korea, for example, the 13/14-year-olds even did better in science, the fewer experiments they did; the case of Korea also showed that it was possible to attain high average achievement levels in science in spite of high average class size.

The fact that certain factors – such as class size or laboratory work – did not consistently prove to have a significant effect on the science achievements does not necessarily mean that they did not have any effect on the quality and outcome of science learning. The tests were mostly restricted to 'paper and pencil' tests and did not include any practical work; it was, therefore, not surprising that the science achievements thus measured did not show any consistent correlation with the amount of practical science activities that students had done[7]. The role of practical activity in science education is further discussed in *Chapter III*.

It should also be noted that although consistent relationships between some factors and student achievement in science were found for a majority of populations, counter-examples almost always appeared. Factors that affect academic performance interact in complex ways and operate differently in different cultures and education systems (see IAEP/ETS, 1992a.b.).

Section 7: Destination of school leavers

Measuring the impact of science education has to go beyond measuring achievement levels. In his reviews of various studies on developed countries, Walberg (1991) noted that it was very difficult to establish associations between academic grades and later scientific distinction or occupational success. The latter clearly depend on a range of factors that are much wider than those normally measured by assessments used to test students in science. Thus, one measure of the impact of science education it might be useful to examine is the flow of students into higher education and the labour market. Ideally planners would like to know:

- How many secondary science students continue into higher education in science-related subjects, how many join

7. Moreover, the analysis conducted tried to factor out contributions to the variance of total score. If a dependant variable showed little variation, e.g. where textbooks were abundant, it would not show up as a significant factor in determining achievement. Perhaps if they were all removed they might!

vocational training programmes of different kinds and how many enter the labour market?

- What sort of studies do science school leavers undertake at higher education level and is a science background required, or not; what is the amount of 'wastage' among science graduates?
- How well do they fare in their studies at higher level and have they been well prepared to continue studying science at higher level?
- How successful are science school leavers in the labour market; do they find a job easily and do they have a good salary, or are they unemployed? Are potential employers of science school leavers and graduates satisfied with their skills, knowledge and attitudes?

Having information on the first and second points allows planners to assess whether the output of the school system is in balance with demand from higher and further education systems and where corrections should be made, if any, in the flows though secondary level. Information on the third and fourth points informs planners on the quality of science education at secondary level as perceived by higher education institutions and the labour market. It provides information on whether there are enough (or too many) science graduates and school leavers being trained considering the jobs available, or whether the outcomes in terms of skills and attitudes developed are really those that are valued by employers.

1.   Flow of science school leavers into higher education

Access to post-secondary science studies

The different countries studied in the IIEP research had different selection systems to enter post-secondary education. In most countries, the selection was done on the basis of a national end-of-secondary examination (France, Burkina Faso, Senegal, Botswana, Kenya, Jordan, Morocco, Malaysia and Papua New Guinea), while in others it was on the basis of national university entrance examinations (Chile, Japan, Korea, Thailand). The selection varied in severity in these examinations. Various sub-sectors of higher education tended to set their own selection criteria. The prestigious institutions, in particular, which played an important role in the education of the scientific élite, applied selection criteria which were a lot stricter than other post-secondary institutions: e.g. the so-called Grandes Ecoles in French-speaking countries – as in

Senegal, Morocco – the prestigious science and engineering faculties in
Japan, Korea, Chile and Thailand, and the Faculty of Medicine almost
everywhere. Another sub-system applied much softer criteria, accepting,
for example, students who had not passed the examination or who did not
have good grades (e.g. the Open University in Thailand). In Argentina,
there appeared to be no selection examination at all, except the one
organized by the Faculty of Medicine.

Proportion of science students entering higher education

Obtaining information on the flow of science students into higher
education is not as straightforward as it might appear. Using cross-
sectional data, it requires that one should be able to assess on one side:
(i) who is a science graduate (as seen in *Section 1* of this chapter, this
information was not always readily available); (ii) how many science
graduates are being trained in public and private institutions; (iii) how
many study on their own or pick up studies after having stopped for a few
years; and, on the other side: (iv) what are all the educational and training
opportunities open to science graduates; (v) what is the intake in these
different university and post-secondary courses, including private ones;
(vi) how many science students go and study abroad with or without a
scholarship, and (vii) how many enter the labour market directly. The
various methodological difficulties that such an exercise entails will be
discussed in *Chapter V*.

In systems inspired by the French tradition, whoever has the
baccalauréat (which is fairly selective, at least in most African countries)
tends to continue at the university or post-secondary level. Science
graduates have wider educational opportunities, higher chances of going
abroad and, hence, higher transition ratios. In Morocco, for example,
98 per cent of all experimental science graduates continued studying at
higher education level. A number of other countries also had a fairly
high proportion of their science graduates entering post-secondary
education: Chile, Argentina, Jordan, Thailand and Papua New Guinea.
Yet, as seen earlier in *Section 1*, Chile and Argentina had serious
problems in filling their science and engineering places at university level
with qualified graduates.

The proportion of science graduates entering higher education was
lower in Korea and Malaysia, but they still had more opportunities than
art graduates. *Table 2.9* below compares estimates of the 1989 output of
Malaysian school leavers likely to be qualified for higher education and
training with the number of places available and filled in 1990. From this
it is clear that science students enjoyed a more favourable selection ratio.

The number of places available in local public institutions was 40 per cent of the estimated output of science students; (the figure for arts students was 19 per cent). Overall, it appeared that places were available for about 70 per cent of science students and 45 per cent of arts students when all sources of supply of opportunities were included. The 1991 cohort of science school leavers was substantially smaller and thus the selection ratio was likely to be even larger. Since science students could enrol in non-science courses, whereas arts students might be precluded from much science-based education and training, selection ratios were probably even more favourable for science students than they appear above. This means basically that almost all science-qualified leavers, who were both able and willing to continue studying in science-based courses, were currently needed to fill the places becoming available.

Table 2.9 Malaysia: Supply and demand for higher education and training

Demand		Supply				Places as a ratio of total school output	
School output		Institutions		Places available			
Science	Arts			Science	Arts	Science	Arts
		Local public only		15 193	17 213	0.40	0.19
		Local private only		2 373	9 494	0.06	0.10
		Total local		17 566	26 707	0.47	0.29
		Overseas		8 866	14 534	0.24	0.16
37 552	91 652	Total		26 432	41 211	0.70	0.45

Source: Sharifah Maimunah and Lewin, 1993.

Type of studies undertaken by science graduates at higher education level

The Moroccan study provided some insights into the type of studies undertaken by science graduates at higher education level and on the amount of wastage. The data were obtained by analysing the files of (nearly) all students, which were kept by the Ministry of Education. It appeared that a vast majority of 1986 secondary science graduates, nearly

90 per cent, did enrol in science faculties, largely in those disciplines for which they had been trained *(Table 2.10)*. Only 4 per cent started studying other subjects, such as economics or business. A year later, however, enrolment in science studies had dropped by 8 per cent, some students having dropped out and others having moved to the economics faculty.

Table 2.10   Follow-up of experimental science secondary-school leavers in further education, Morocco, June 1988

June 1986	Areas of studies at post secon-dary level	1986/1987			1987/1988		
		September 1986	June 1987		June 1988		
			Enrolled	Passed exams	Year of study	Enrolled	Passed exams
	Physics-Chemistry	7 031 *41.4%*	6 985	1 207	I	5 197	1 601
					II	1 179	493
	Biology-Geology	6 852 *40.4%*	6 713	1 879	I	4 353	1 913
					II	1 866	1 158
	Maths-Physics	16 *0.1%*	22	5	I	nd	nd
					II	nd	nd
A-level holder	Technical training and technology	53 *0.3%*	95	42	I	30	35
		.			II	42	36
Exp. Sciences June 1986	Medical training	755 *4.4%*	800	541	I	418	349
					II	540	439
16 947	Economics-Management	629 *3.7%*	715	259	I	805	307
100%					II	259	180
	Literary and legal training	113 *0.7%*	126	16	I	164	30
					II	16	12
	Other	55 *0.3%*	77	28	I	92	41
					II	28	25
	Total	15 504 *91.4%*	15 533			14 998	

Source: Caillods *et al.* (1997).

The analysis of the same data for the 1990 graduates, indicated that fewer experimental science graduates enrolled in science faculties (65 per cent instead of 86 per cent: the wastage was much more important). But there were 20 per cent enrolling in the economics faculty. The reasons for this shift were to be found in: the deterioration of employment prospects for science graduates on the labour market, and possibly in the change in the language of instruction at secondary level[8]. Many students were said to be afraid of not understanding the lectures at higher education level. The usefulness of monitoring such development is clear. Should such a trend be confirmed over several years it might be an indication to the Ministry that either the content of the experimental science stream has to be modified or that the number of pupils guided into it ought to be reduced in favour of the economic stream which is cheaper to organize.

Performance of science graduates at higher education level

The same 1986 and 1990 cohorts of secondary school graduates in science have been followed for a number of years at university level in Morocco. The data obtained show that, for example, the repetition rate in the first year of higher education was extremely high (between 60 and 72 per cent) for the experimental science graduates, higher for those students who studied physics and chemistry than for those who studied biology and geology, and much higher than the repetition rate of the mathematics/physics stream graduates.

Of course, such figures had to be interpreted carefully before any conclusions on the quality of science studies at secondary level could be drawn. High repetition rates might indeed reflect an insufficient level of preparation of the science students when they entered university to study that subject, but also their level of understanding of the medium of instruction (French). They might reflect the very poor teaching conditions prevailing at the university during the first year of study as well as a policy of screening students in a country where there was no selection at university entrance. As for the better performance of mathematics/physics graduates, it could probably be explained by the better level of preparation of this group of students as well as by the fact that they had been highly selected. The sort of information which was useful for the planners was that science graduates had better results than art graduates,

---

8. In 1986, students were taught in French both in secondary and in higher education; in 1990, science was taught in Arabic in secondary school and in French at the university.

and that, provided a special effort was made to help first-year students, the policy regarding the change of language at higher-education level did not have any negative effect on students' performance – on the contrary. Another interesting finding was the fact that science secondary graduates had a lower level of performance in the economics faculty than those studying economics; although this could be expected, the latter being better prepared as from secondary education, this might be an indication that experimental science graduates were too narrowly specialized, particularly if future jobs were more likely to be in the area of applied science rather than in pure science.

2.   Employment opportunities for science school leavers

It is generally assumed by planners, administrators, students and parents alike that science graduates have better employment opportunities than art graduates. This is related to the particular contribution of science and technology to growth and development. By implication, this should appear in labour market signals that provide higher salaries to science graduates than to those in arts and humanities. Whether this is true or not in particular labour markets needs to be established.

It was virtually impossible to evaluate information on the employment and work opportunities for science-trained secondary school leavers. What little there was at a country level was piecemeal, and unreliable. Tracer studies were rarely conducted. Though there was evidence of growing unemployment of lower secondary school leavers in many countries, this could hardly be attributed to the way science was taught. At upper secondary level, it was not possible to establish whether science-qualified school leavers were better or worse off than their peers on the labour market. Many of them continued into higher education anyway. Even in the developed countries where tracer studies are conducted, science-based school leavers at secondary level are not separately identified from other students: it is therefore not possible to compare their job opportunities to those of other secondary school leavers. Even if it can be shown that science graduates are more attractive to employers, it would be difficult to ascertain whether this is because of the knowledge and skills they have acquired or simply because they come from a selected group of students with higher than average ability.

There is slightly more information on the employment opportunities of higher education science graduates. Numerous reports written at national or regional level in, or on, advanced countries, newly industrialized countries or low and middle-income countries consistently stress the need to increase the number of science and technology graduates to support economic development. *Chapter I* argues that it is not enough to produce large numbers of science and technology graduates for development to occur. These graduates have to be engaged in productive employment. Otherwise, they will end up increasing the number of educated unemployed who have been trained at very high cost, or feeding the brain drain. Some of the data collected on employment and income opportunities of higher education science graduates are reviewed below.

Employment opportunities of science graduates

In most countries, science and technology graduates are primarily employed in the public sector (public administration and para-statals). The biggest employer of science-trained graduates is often the Ministry of Education with the teaching profession. This is the case in the most advanced market economies, where not only teachers, but also researchers (physicists, chemists and engineers) are employed by the public sector. In developing countries, possible jobs for science graduates are mostly concentrated in the public sector, as the private sector is usually much less developed. As a result, employment opportunities for science graduates very much depend on job openings in the public sector, and are often linked to fluctuations in the recruitment of teachers.

From the IIEP survey it appeared that employment and remuneration prospects were reasonably good in countries which enjoyed rapid industrial growth and where the introduction of new production technologies was forcing businesses to recruit more and more employees with an advanced scientific background – this was the case in Korea, even though overproduction of science graduates had led to some unemployment, and in Thailand, France, and Japan. The situation varied between types of graduates. In most OECD countries, engineering graduates are amongst the most advantaged of all, having minimum initial unemployment, a short job search period and, as shown below, initial wages among the highest. The picture is more mixed for other scientists: while in a few countries they might enjoy similar employment conditions to engineers, in others, their employment situation is similar to that of an average university graduate (OECD, 1993).

In Africa and Latin America, the economic recession of the 1980s affected the employment prospects of science graduates. The structural adjustment programmes which had been introduced in many of these countries considerably slowed down, or stopped, recruitment into the public sector. This led to unemployment of science graduates, even if their job prospects were still better than those of other graduates. In the mid-1990s, it became common for low-income countries to suffer at the same time from shortages of experienced scientists, engineers and technicians, despite large-scale investment over a lengthy period, and from graduate unemployment. Graduates may prefer temporary unemployment to taking jobs that do not correspond to their career aspirations (especially the case with pure science graduates, for whom research jobs may be in short supply); and the better qualified may migrate.

In developing countries, published data on employment opportunities and wage levels of science graduates, even those from higher education, are scarce and incomplete. In IIEP's two case studies, whatever studies and statistics existed have been reviewed and this analysis has been complemented by a review of newspaper job advertisements and with some interviews with a small selection of key informants: admission officers, academics, and employers. The case studies are illustrative of two contrasting contexts – a situation where science graduates were in high demand thanks to the industrialization process (Malaysia); a case where the shortage of science graduates had been replaced by a relative oversupply (Morocco), as the economy and employment had not yet recovered.

In the late 1980s, Malaysia went through a short period of recession. As government slowed down its recruitment, reducing access to teacher training in particular, science graduates were said to remain more unemployed than others. This was due to their alleged poor communication skills, notably in English, and insufficient adaptability. This was not entirely confirmed by employers' interviews, but there was some suggestive evidence that employers (and institutions of higher learning) were not satisfied with the quality of science graduates. In 1990, the analysis of available labour market information identified substantial numbers of vacancies for science-qualified school leavers. Plans for economic growth – the Malaysian economy is targeted to grow by an average of 7 per cent per annum until the year 2000 – have created concern that there might not be enough science-qualified students. A critical shortfall has been identified in civil, mechanical, electrical and electronical engineering, as well as in some fields such as the computer and electronics industry. The issue has become how to attract good students to science studies.

Morocco was a different case in the 1970s and 1980s. The policy was to expand greatly the number of science and technology trained graduates, to support a strategy of economic development and nationalize the cadres in general and the teaching force in particular. This had led to a rapid increase in science enrolments at the university and at upper secondary level. In 1993, all teachers were Moroccan and the Ministry of Education had considerably reduced the number of teachers recruited every year. Meanwhile, expectations that a rising number of science graduates would be recruited by commerce and industry had not been confirmed. This led to a drastic decline in the employment opportunities of science school leavers. While job prospects remained very good for engineers and graduates of the highly selective schools and specialized Institutes – many of whom were recruited before they even finished their studies –, graduates of pure science faculties (biology, geology, physics and chemistry), who expected to become teachers, seemed to have very few job openings. It was reported that many of them had great difficulty in finding other work, either because they were not well prepared for other occupations or because they did not accept other offers. It was also frequently argued that they appeared less entrepreneurial and more resistant than other graduates to the idea of establishing their own business. This could be due to the costs of opening a business (in the case of medical doctors, for example) but these costs are often overestimated. When there is effective demand, capital is likely to be available. It is more likely that this was related to the high aspirations that such graduates had and to the fact that they might prefer waiting until a government post opened.

Whether science graduates are really more risk-averse than others (which is still to be tested) or not, there is a case in Morocco, as in many African countries, for reviewing the content and the type of courses offered. Science studies might fail to teach the skills that they are supposed to teach – scientific thinking, a creative and critical sense and problem-solving skills. The content of the science courses might also have to be changed to be more oriented to the countries' needs and economic circumstances. Science and technology manpower has to be trained to be more versatile, and more able to adapt to changing situations.

Science graduates' salaries

According to the economic reasoning, a good indication of whether or not some graduates are in high demand on the labour market should be derived from the salary structure. If there is a real deficit of science graduates on the market, this should be reflected in their salaries; at the same time, high salaries would provide an incentive for qualified students to study science.

Some of the data accumulated on developed countries (*Table 2.11*) show that salary levels differ widely according to whether students have specialized in applied science or pure science. Generally speaking, engineers, computer scientists and medical doctors receive much higher salaries than natural scientists.

Table 2.11.   Starting salaries of graduates of different specialities in the
USA and France

United States of America Starting salary (US$)		
Graduates	US$ 1989(a)	US$ 1993(b)
Medicine	$60 000	
Engineering	$50 775	
MBA	$43 883	
Physics and chemistry	$43 215	
Bachelors		
Engineering	$30 611	$30 500
Computer science	$28 296	$31 783
Maths and physics	$27 000	$28 221
Business Administration	$22 274	$24 584
Humanities	$20 243	$23 361

France (recruitment salary in industry and commerce in French Francs)	
Top engineering school	more than 210 000FF
Other engineering schools	180 000 - 190 000FF
Science Phds	200 000 - 210 000FF
MSc Computer	195 000FF
Top Business Schools	200 000 - 210 000FF
Average Business School	177 000FF
MA Finance	195 000FF
MA Law	190 000FF
Bachelor law	175 000FF
Economics	175 000FF

Source: (a) USA Department of Commerce.   (b) USA College Placement Office.
France: Les dossiers de l'expansion, avril 1995, No.498.

The former receive salaries comparable to those of managers and executives, while the latter often fare no better than arts graduates. Many natural scientists are employed in the teaching professions where salaries are generally low. Public sector salary structures are often not a good guide to labour market signals since they are often slow to change in response to shortages.

No similar data are available on developing countries, but a recent study on civil service salary scales in Africa (Robinson, 1990) shows that scientific and technical personnel may sometimes be placed on a separate, slightly higher salary scale, than other civil servants or that they may be placed on the same salary scales but start at a higher point. It appears, however, that higher-level posts in the administrative class are often occupied by non-science graduates. The picture that emerges of this structure of rewards does not suggest that there is convincing evidence that typical patterns of remuneration are sufficient either to encourage African students to study science or for African science-based professionals to remain in the country. Internal pressures to maintain the civil service salary scales make it difficult for government to offer much higher salaries to scientists, and to science teachers in particular.

A study carried out in South Asia also showed that scientists and engineers in the civil service did not receive higher salaries than general administrators. Entry points on the salary scale were the same and promotion opportunities were fewer. Also top posts in the technical and professional service were often graded lower than those in the administrative service. As Chew (1992) says, the higher status and salary of generalist administrators transmits the wrong signals to secondary and tertiary students, and reinforces a non-applied science white-collar administrative mentality.

When planning intakes into science studies, planners should indeed take into consideration the salary structures. If they find that there are no income advantages to studying science, and/or the prospects of being employed are no better than for others, this suggests several possibilities which have to be carefully evaluated:

- it may indeed be that there is a surplus of supply over demand, or that, though the quantity of output is adequate, the quality is low or the content irrelevant. Planners should however know that overall surpluses may mask acute shortages in some specialities;
- it may also be the case that labour market signals are distorted – where most modern-sector employment is in government administrative bureaucracies this is almost inevitable. In order to encourage students to study science, incentives other than monetary ones may have to be considered;

- effective demand for scientists and engineers may appear weak
  because employment practices in the private sector discriminate
  against nationals.

In conclusion

In this chapter, an attempt has been made to establish the state of
science education in a variety of developing countries. In order to do so,
an analysis has been made of the coverage of science education at
different levels, the organization of the curriculum, the teaching and
learning conditions, the teaching practices and the cost of science
education, before discussing the impact of the effort made on educational
achievement (what pupils have actually learnt) and on the destination of
school leavers.

The state of science education, of course, varies a great deal between
countries and sometimes even more within countries. Most governments
have made tremendous efforts to increase the participation of their young
population in science education. The organization and curricula of science
education and the forms and degrees of specialization in science subjects
continue to differ between countries and this reflects their respective
history and culture. Certain conditions of science education provision –
such as the qualification level of science teachers – have improved almost
everywhere; most African countries, however, continue to face a lack of
financial and human resources, affecting the conditions of science
teaching and learning. The research also provides some evidence that
even in countries where essential resources appear to be available, a lot
remains to be done to improve the quality of education and student
achievement in science.

Participation in science studies has increased everywhere, but
disparities in access at Grades 7 to 9, where science is taught in a more
structured way, remain very large. Patterns of science achievement also
vary significantly among countries and, within each country between
different geographical areas, schools, etc., as available studies on science
achievement show. Existing findings on the factors affecting science
achievement suggest that the latter depends in particular on curriculum
content, the amount of contact hours devoted to science, availability and
quality of textbooks, and the level of subject knowledge and subject-
related pedagogical skills of teachers. It was found that in practice – even
in cases where the official curriculum sets aside substantial amounts of
time for science – the quality of secondary science education provided
tended to be seriously affected by insufficient and inadequate materials
and shortcomings in the training, support and career prospects of science
teachers. Examination pressure, curriculum overload or lack of a precise
and operational curriculum and assessment objectives, appeared to be
further factors impeding effective science teaching.

Measures should be introduced to increase the appeal of secondary
science education. Unless this is done, it will neither be possible to

improve the quality and outcome nor to match more evenly the supply and demand of science graduates. Human resource planners may indeed find that in spite of relatively favourable employment prospects for science graduates, stronger incentives, both monetary and non monetary, will have to be considered in order to encourage students to study this subject and enable them to succeed. This is one of the several issues to be tackled in planning science education provision.

Chapter III
Main issues in planning science education

Educational planners and policy-makers have to decide as to how science education should be provided at secondary level. Their choices have to be based on existing and foreseeable economic constraints as well as their respective country's needs in human resources. It is crucial to know the number of students who should be directed towards specialized science programmes at this level, the kinds of specialization which should be taken and the criteria and procedures to be applied to the selection and tracking of students into such programmes. How can science be made more attractive to all secondary students, in particular to those student groups (for example, girls and rural children) who are less likely to study science beyond the lower secondary cycle? What kind of curriculum changes (particularly the use of local languages) need to be introduced to facilitate science education? How much science should be taught at basic education level; to what extent should components such as technological and environmental education be included; what type and amount of practical activities would enhance the quality and relevance of secondary science education? How can the training of science teachers be improved? How can student assessment in science be made more effective to encompass the major aims of secondary science education?

Section 1: Specialization

There are two main reasons in favour of specialization. First, it may be the most cost-effective way of deploying scarce resources for secondary science education – if there is no realistic prospect of adequately resourcing all schools, it may be better to provide well-founded facilities for a selected minority. Second, some level of specialization may be the best way of responding to the range of abilities and interests that secondary school students manifest – it allows those with most ability to study in depth, permits student preferences for science and other subjects to be expressed, and does not preclude all students from studying some science. Some form of specialization in science is therefore almost inevitable, and in all but one of the countries studied it occurs in some form. In the exception (Kenya), it is widely recognized that informal specialization results from the different resource endowments of different types of schools. This means that science is taught most effectively in a minority of secondary schools which attract many of the best science students.

Specialization can take many forms. Three different types of curriculum-based specialization were identified in *Chapter II*, where the number of subjects, the time allocated to science, and the differences in allocations between science and non-science students vary. In addition, some countries provide completely separate secondary level science schools which select the best students and provide them with the best teachers and facilities. This option is discussed further in *Chapter IV*.

Here, four issues are discussed on which educational planners have to make decisions related to specialization in science. These are:

(i)   at what stage of secondary schooling should selection for specialization occur?
(ii) how many secondary students should specialize and to what depth?
(iii) how far should students determine choice of specialization?
(iv) what forms of orientation and guidance should be provided?

1.   Early versus late selection

Selection occurs most frequently at the end of lower secondary. In some cases (especially where special science schools exist), selection may occur earlier at the end of primary school. Longitudinal studies on pupil achievement in science (reviewed in Walberg, 1991) do suggest that those who do well early tend also to do well later, that success at a young age provides motivation to continue to achieve, and that mathematical and scientific skills and attitudes benefit from early development. Measured science achievements of secondary students are often reasonable predictors of grades on standardized achievement tests at university level. By age 12, correlations are beginning to be significant in those (developed country) populations where studies have been completed. It is argued that these correlations may be particularly strong in science and mathematics because these subjects are 'hierarchical' – advanced topics are difficult to learn if the basics are not mastered. A conclusion that could be drawn from this is that it may be possible to identify those who are 'capable in science' early provided the assessment of science achievements is technically adequate.

Some caution is needed in extrapolating these results to policy on early selection in developing countries. Most of the correlation studies have been undertaken on school populations in industrialized countries. Where school provision is more varied in quality it may be the schools rather than students that are tested by selection examinations. If correlations are examined amongst those who are successful in reaching higher education, the only confident causality is that those who do well at higher levels also did well at lower levels and that within the group a ranking of performance is maintained. It remains unknown what those excluded might have achieved (if, for example, they were 'late developers') since they are not included in the analysis. Not surprisingly, the older the population assessed, the higher the correlations are, and the more reliable. The younger selection occurs the more likely that reasoning patterns are unstable and in a transitionary phase between child and adult. Students acquire cognitive capabilities at uneven rates and at different ages; those who experience rapid development in their late teenage years may be disadvantaged if selection occurs early.

There are other serious arguments against over-emphasis on early specialization in science. There is a fairly compelling case that some priority should be given to 'Science for all' at the primary and lower secondary levels. In most developing countries, these levels are the last at which there is a chance to acquire the rudiments of scientific literacy. Quality basic education must include knowledge and skills in science as a central area of knowing and this may be at least as important for national development as special provision for the most talented. Quality basic science education is also important because science learning and its outcomes at the subsequent levels depend to a large extent on the foundations established over the first eight or nine years. Both for those entering the labour market and for those continuing to study, an adequate grounding in science is a prerequisite.

Another argument against early specialization is made by those concerned about unequal participation in science studies by gender. Research in the United Kingdom and Australia suggests strongly that where secondary science courses are made optional, girls disproportionately tend to opt out of science, or concentrate only on life science options. IIEP's research supports this conclusion – thus in Malaysia, girls were far less well disposed to studying physical sciences than boys and, when choices were given, tended to opt for biology or non-science subjects (Sharifah Maimunah and Lewin, 1993). They also generally score less on science tests and are therefore more likely to be excluded from the science streams. One consequence of such a phenomenon is continued under-representation of girls in science-based courses at higher levels in most of the countries studied, reflecting a general pattern (Harding, 1992). Where girls continue studying science their performance may be as good as that of boys. Thus, it may be both inequitable and inefficient to allow girls to opt out of science as a result of early specialization.

However, allowing for early selection and specialization in science may not be incompatible with mandatory science for all throughout the greater part of secondary education. The disadvantage of specialized science is that where it absorbs most of the resources available those in other streams tend to suffer from a diluted programme. In such a case, the cost of the selection of an élite for specialized science conflicts with the goal of science for all.

Decisions as to when selection should take place have to balance the potential gains for the selected groups against the educational disadvantages for the others (Lewin, 1992). A solution has to be found that does not entirely deprive the majority of students of science education at secondary school level. Equity and efficiency may suggest the delaying of specialization until students are considered to have a degree of maturity and have had a reasonable chance to acquire basic skills at a level where schooling is generally thought to be of an acceptable quality (i.e. at the end of lower secondary: grade 9 or 10). The loss to those who learn fast may be less than the advantages gained by extending scientific literacy up to a certain age and then selecting for specialization from a larger pool in a more reliable manner. However, where school quality is poor and access is very restricted, selection and de facto specialization will occur on bases which often have little to do with drawing from the national pool of talent of all children of school age.[1] Under such circumstances, early selection at the beginning of lower secondary from as large a pool of pupils as possible may be the best option.

2. Forms of specialization

In some countries, secondary students specialize through enrolment in a science stream; in others, students can choose between two or more

---

1.　In every country there are élite institutions where science is taught in a more appropriate way. The question is: how are children selected to enter such schools?

science streams, each offering a different menu – i.e. balancing the different science and other subjects in different ways.

Another way of introducing some specialization is through the choice of options. Specialization is made 'à la carte' and students can choose – usually within certain limits – the number and composition of optional courses or electives in science. Korea, Thailand and Japan have adopted this kind of optional system in which the number and composition of electives may vary; Malaysia is beginning to adopt this at upper secondary level. As with all option systems, choice will be constrained for students by what is actually available in the institutions they attend and this may be less than the full range of possibilities.

The degree of specialization is a function of both the number of subjects studied and the amount of time allocated to them, which gives an indication of depth. Where two or three science subjects (not including mathematics) are studied, making up more than about 25 per cent of curriculum time, specialization is likely to be high. The same time allocation spread across a wider range of science-based subjects offers less depth but still a high general level of specialization in science. Where time allocations are much smaller – say 15 per cent or less – specialization in science is likely to be weak. In Japan and Korea, students specializing in science are reported to spend particularly large amounts of time on the learning of science (see *Chapter II*). This is a product of the option system, which allows additional science subjects to be added to the compulsory core curriculum. Furthermore, in these countries and everywhere where science is popular, much time may be spent out of formal school on private tuition in science-related subjects.

Programmes which have very high levels of specialization restrict the array of possible further educational and professional choices. Narrowly trained students are less flexible than those who have followed broader-based programmes when it comes to seeking further education or employment outside directly relevant specialities. This may be the case if the science education that has been experienced is strongly academic and theoretical and the opportunities are applied and technologically orientated. Generally, opportunities for applied science-based graduates appear much more frequently than those for academic and research-based scientists. Also, in many developing economies it appears that polyvalent skills are being preferred to specialization in large parts of the labour market.

3. How many should specialize?

There is no single answer to the question: 'what should the appropriate proportion of secondary science students be?' This depends on a wide variety of factors which include the proportion of the relevant age group enrolled at the upper secondary level, the demand in the labour market and from higher levels of education and training, the stock of science-qualified school leavers, and the historic patterns of curriculum organization. Practice on when some form of specialization is introduced varies between countries, as noted in *Chapter II*. In some, one-quarter or less of those enrolled in general upper secondary education specialize in science. In others, more than half of the students in general secondary education follow some specialized training in science.

In countries like France, Korea and Thailand, the relatively high proportions of secondary students specializing in science may be justified by the dynamics of economic development which generate a large number of science and technology-based jobs. Malaysia and Chile also have dynamic economies which share many similar characteristics which might be thought to call for a large proportion of science enrolment. However, they record a relatively low percentage of secondary students specializing in science, a problem we shall be returning to in the next section. On the other hand, countries like Senegal, Burkina Faso and, to some extent, Morocco have high proportions of secondary students specializing in science, an emphasis which may be partly explained by their low enrolment ratios at upper secondary level and the high labour market demand for localization of expatriates. In the late 1980s in these countries, the market demand for science-trained staff (with the exception of science teachers) stagnated, and the respective government's policy may have to be reviewed in this light.

Where there is evidence of imbalance – too few places being taken up despite a strong demand by employers – and signs of unequal participation (by girls, rural pupils, etc.), this is likely to be a convincing case for an increase in the numbers studying science. Even where employment growth in science-related jobs is sluggish, underlying demand may be strong (temporary economic downturns may suppress demand in exaggerated ways, localization may still be needed). When, on the other hand, unemployment amongst science-qualified school leavers and graduates arises from unrealistically narrow job expectations (e.g. graduates are unwilling to follow careers in anything but academic or research occupations), there may be a case for reducing the proportion of students specializing in science and redeploying the resources for the benefit of all. In other words, where demand from higher levels and the labour market is weak, expanded specialized provision may not be a priority. Conversely, where demand is strong, enhanced output may be desirable.

How this is best achieved depends on the conditions under which policy can be implemented. Policies on the numbers specializing in science have to depend on the resources available and the capacity for effective implementation. Severe resource constraints require concentration of effort (and substantial selection and specialization); richer systems may be able to afford more generous provision that does not require such concentration of effort.

The argument is made in *Chapter IV* that effective science education is not necessarily much more costly than other subjects. It all depends on the nature of what is provided and the educational goals to which it is directed. Several countries attain very high levels of science achievement without extensive use of expensive practical facilities (e.g. Korea). Though there is a prevailing view that genuine science education can only take place through experimental hands-on activities in fully equipped laboratories, the empirical basis for this is weak (as discussed later). If, at the same time, a margin for manoeuvre exists for enlarging the size of the science classes, it may be possible to expand enrolment in science programmes for all without increasing the expenditure on the provision of science in an unsustainable way.

A final point to be taken into consideration is that it is not easy in practice, and probably not desirable, to plan and implement a policy of specialization without recognizing student preferences and socio-psychological aspects of learning as the cases of Morocco and Malaysia illustrate. Children obviously develop at different rates, in different directions, and acquire different tastes and skills. Typically, attitudes to science deteriorate during the secondary school years and not all secondary students are interested in, nor achieve well in science or in mathematics. Effective learning requires motivation as well as matching the cognitive demand of curricula to realistic appraisals of what students are capable of doing. The organization of (upper) secondary education has to take these differences into account if learning at this level is to be meaningful. Somehow, individual characteristics of students have to be reconciled with judgements of rational investment of national resources in provision of different types. Individual preferences will not produce ideal patterns of investment, and indicative planning may attempt to fit too many round pegs into square holes.

The issue of the proportion of students who are selected for specialized courses at secondary level is discussed further in *Chapter IV*, where the advantages and disadvantages of special science schools are discussed. At this stage, it should be stressed that the amount of specialization, as a general issue, should be separated from the narrow question of how best to cater for the needs of a future scientific and technological élite. It may be that intensive and accelerated programmes are appropriate for small minorities with special talents who can be taught in specially selected groups and/or in special science-based schools. Interestingly, the science achievement of the most able students does not seem to depend on the overall proportion who specialize at higher levels in the school system (Postlethwaite and Wiley, 1992).

4.   How should the selection be made?

Practice differs on the extent to which allocation to science streams is directive or involves an element of choice. The two extremes are where students are directed on the basis of examination results, with the highest scoring expected to study science, and where students have a completely free choice of subjects/streams to enrol in, providing they satisfy a minimum entry criterion. Allocation policy is used in some cases to reduce disparities in participation between different regions or different ethnic groups. Where schools are single sex and the number of places for girls and boys differs at higher levels in the school system as a whole, this may be interpreted as reflecting policy that favours one or other group.

Francophone systems tend to allocate places on the basis of recommendations from the class council (the teachers who have taught the students) and the number of places available (as in France, Burkina Faso, Senegal, and Morocco). Students and parents are informed about the range of possibilities and students are asked for their preferences, but the decision is usually not theirs. Mediators may be called in where there is disagreement between teachers and parents about a pupil's future career.

Morocco has been following this type of directive allocation into secondary science while seeking to reduce the existing disparities of access to upper secondary education. A kind of quota system is used whereby the Ministry ensures that 40 per cent of lower secondary completers from each school are offered places in upper secondary. Standards vary between schools: this practice therefore ensures representation from all regions and groups. It is not strictly meritocratic since actual levels of achievement on entry will vary widely depending on the school attended. In other words, in low-achieving schools, students will be selected who would not be eligible in high-achieving schools. Since schools are widely dispersed geographically, rural students are not excluded from further study by lower levels of achievement. As a result, the percentage of upper secondary students specializing in science in Morocco varies relatively little (between 40 and 55 per cent) among the different provinces (*Académies*) of the country. The result of this policy is a very heterogeneous school population, with a much wider variation in ability than would be the case if selection was purely meritocratic. Teaching is more difficult: more than 87 per cent of senior secondary science teachers consider that the students entering the science stream have not been properly selected; nearly half find that the levels of science knowledge and skills amongst their students are very heterogeneous and only 39 per cent of them think that most of their students are actually 'interested' in the subject matter (Caillods, Göttelmann-Duret et al., 1997).

In most other countries covered by the IIEP project, the allocation of secondary students into science is less directed. In Malaysia, option choice at upper secondary has changed from a directed system which placed the academically most able in the science stream, to one where student and parental choice is weighed heavily, assuming a minimum level of achievement is reached. This is beginning to change the pattern of those who choose science. It may result in a more uneven spread between groups since some are less predisposed to study science than others. Choice has been expanded and a much wider combination of options is now permitted, but actual choice is restricted by the options that schools offer which, in some cases, particularly in rural areas, will be constrained by facilities and staffing.

In Kenya, where options are nominally available in all schools, differences in the quality of provision have similar effects on who is ultimately selected. A significant number of secondary schools do not offer the three separate science course option. Only a restricted number of fully funded government schools offer such an option. Thus students in some regions and in many rural areas tend to have reduced opportunities of access to this more intensive science training. Where the opportunity to specialize is restricted geographically or otherwise, science students will be drawn from a narrower range of backgrounds than where the possibility of choice is universal. To avoid large disparities, the Kenyan Government has introduced a regional quota system for selection into fully government-aided secondary schools.

Malaysia established special science schools for Bumiputra students and Nigeria introduced a similar initiative with a view to ameliorating indigenous under-representation among secondary (and post-secondary) science graduates . These two specific experiences – namely the Science Schools in Kano State in Nigeria (see Adamu, 1992) and the MARA and Science Residential Schools in Malaysia, will be discussed in more detail in *Chapter IV*. In both cases, these schools succeeded in increasing the number of secondary science graduates belonging to the target population who subsequently enrol in science – or technology-based further or higher education courses.

In some countries, each secondary school decides on levels of student recruitment, on the parts of the curriculum (options or electives) to be provided, and on who gets selected into science courses (e.g. Chile). In such cases, regional disparities in the participation and the quality of the education provided may be particularly acute.

All strategies have their drawbacks. Regional (as well as school-based) quota systems may contribute to equalizing pupils' opportunities as regards access to secondary (science) courses. Without special support measures those selected with a quota may remain at a disadvantage and may subsequently underachieve. On the other hand, relatively high achievers in competitive catchment areas will be excluded in favour of those with less demonstrated achievement. But meritocratic selection alone will tend to over-reward those from favoured educational backgrounds. School-based quotas are probably the least attractive option, unless at the very least, monitoring examinations are used to ensure that minimum standards have been reached. Regional quotas are more attractive, especially if they can be coupled with measures to limit the variation in standards between regions.

5. Orientation and selection into science

Whether specialized science streams or programmes recruit the pupils who are actually the most capable and motivated in science depends on several factors. Most obviously, the method of selection needs to be technically sound (unreliable instruments of low content and predictive validity will not succeed in identifying the most able). If entrants to the selection process are already filtered (by the type of school they attend or its location and resources), the pool from which selection is made will be limited; and where advice and information are in short supply, local prejudices for or against science studies may have a disproportionate influence. Aspects of the design of assessment instruments are discussed elsewhere (see *Section 5*). We turn below to some remarks on the orientation of students.

All the countries reviewed have some form of guidance for students in choosing options and may provide advance information on whether to take selection tests where these are optional. In reality, the quality and type of guidance provided differs very much from one context to another. Most commonly, it is provided by classroom teachers with no special background in diagnostic assessment and with very varied degrees of knowledge of opportunities related to the study of science. Where advice is available from those with a science background, most commonly it is from science teachers who themselves have not worked outside the education system. In the countries on which IIEP conducted case studies, a very limited amount of material appears to be available in schools which provides encouragement to study science, information on career choices and further study options that depend on science, and indications of the prospects of being successful in taking up opportunities.

For obvious resourced-based reasons, more is invested in guidance and counselling activities in relatively rich countries than in those where the most basic provision is still problematic and secondary enrolment rates are small. Some countries have made considerable progress with initiatives to diffuse information and encourage more students to study science. In Botswana, one hour per week is devoted to 'career guidance' in the first year of upper secondary (the year preceding a decision on pupils' specialization) and special extra-curricular initiatives such as the 'Science Road Show' have been organized to attract young people to specialization in science courses at upper and post-secondary levels. In Thailand, 'guidance teachers' have been nominated in all secondary schools who are supposed to help pupils make their choice of options. However, in almost all countries surveyed, there were indications of dissatisfaction with the availability of guidance at the stage of transition from secondary to higher education.

The difficulties should not be underestimated. Middle-year teenagers do not typically have well-formed and stable preferences; judgemental and psychometric assessment of career choices is not demonstrably reliable; models derived from European cultures may not be appropriate for projection onto students from other cultures with a different hierarchy of values. 'What kind of job do I want?' has a different significance to those seeking to exit subsistence economies than it does for those who can reasonably expect secure employment. 'Who am I?' may be a more important question to some adolescents than 'what can I be?' depending on whether they are part of more individualistic or collective societies. The role of 'significant others' in decisions of educational and occupational choice demonstrably varies across cultures (Little et al., 1987).

Effective guidance may help increase the number of pupils interested in science and willing to specialize in science subjects; certain groups – girls and rural children in particular – may be encouraged to opt for science at upper secondary level if negative stereotypes are contested. The effect will probably be small unless these kinds of activities are complemented by sympathetic changes in the selection process which are more accommodating to groups currently excluded. Improving the numbers who choose to study science also needs to be accompanied by curriculum changes that ensure that those who are selected can benefit from the course that they follow and are adequately supported.

A few concluding remarks

(i)   Selection into specialized science below upper secondary is generally undesirable. It is likely to be unreliable, it may exclude late developers and educationally underprivileged groups, it is not clearly justified by the balance of costs and benefits.

(ii) Selection methods need to satisfy criteria of equity and efficiency, and encourage improved quality. Decision-makers will also have to estimate carefully the cost-effectiveness of different strategies (national assessment tests, school-based selection, quota systems, etc.) aimed at defined goals.

(iii) There is a need to improve the assessment procedures on which selection into secondary science is based. Where this is the result of unmoderated school-based assessments, the results are unlikely to be technically robust. Where standardized tests are used, efforts need to be made to improve the reliability and validity of test items and to follow students through to higher grades to establish the level of predictive validity. If this is low the assessment systems should be refined.

(iv) Planners and policy-makers have to strike a balance between more or less directive methods of regulating access to specialized secondary science programmes. Individual preferences have to be reconciled with national needs for science-trained staff. Where demand for science places is weak, and labour market signals strong, more adequate information and guidance systems may help to improve participation.

Section 2:   How to attract students to science?

In many countries, policy-makers and planners have been, or still are, concerned with attracting more students to science. The production of secondary and post-secondary science graduates has not always been able to keep pace. Shortages of science-trained graduates are all the more critical where expatriates still hold a large number of technical and scientific positions or where, as in Malaysia, Thailand and Korea, high rates of economic growth and industrialization have led to a rapid increase in the number of scientists, engineers and middle- level science-trained professionals required. The number of graduates with a science background needed may also be growing as a result of rapidly expanding enrolments at secondary level which generate a need for science teachers.

Shortages of science graduates may be the result of supply-side problems (i.e. due to the organization of the curriculum and possibly the insufficient specialization of students, as discussed above) or it may be due to a demand problem on the students' side. The factors that influence the supply of qualified science school leavers are country specific, but there may be some recurrent patterns. Science is widely considered to be a difficult subject, which is more difficult to master than some other areas of the curriculum. It may require more sustained effort and intellectual application than other areas. It may also be true that in some countries levels of achievement are set such that pass rates in science are lower than for other subjects, thus making it more difficult for individual students to gain high grades. Attitudes to the study of science are likely to be influenced by the enthusiasm and commitment with which it is taught. Where there are few successful role models encouraging students to study science it will be difficult to attract them to this subject. Perceptions of the labour market may also be influential. If science graduates are not particularly well rewarded in relation to those with other subject specialities, the incentives to persist may be low.

Below are discussed some of the dimensions which influence participation rates in science.

1. Turning away from science

During the 1980s there was a decline in the proportion or number of pupils specializing in science in most of the countries in the study. (This is true for Burkina Faso, Jordan, Malaysia, Morocco, Papua New Guinea, Senegal and Thailand). In most cases, the decrease in science enrolments still persisted, despite government efforts to reverse this trend.

The two in-depth studies carried out in Malaysia and Morocco provided some insight into the problem of attracting (more) pupils to science. In Malaysia, the transition rate into science dropped from 20 to 15 per cent between 1985 and 1990 and, in Morocco, from 23 per cent to 18 per cent. An overall decrease in the proportion of senior secondary students specializing in science occurred in both cases: output in Malaysia was approaching levels insufficient to fill science places at higher levels. A trend away from science was also noted in Chile. The number of students who registered for the specific aptitude test (PCE) in biology, chemistry and physics[2] declined drastically between 1980 and 1991: a 64 per cent drop in biology and an 85 per cent drop in physics (Schakmann, Zepeda and Toro, 1992).

The reasons for such a decline are not altogether clear in Chile. In Malaysia and Morocco, however, perceptions of labour market opportunities were probably a factor in these changes, as discussed in *Chapter II, Section 7*. The (slightly) better job prospects for science graduates and the relatively low market rewards in terms of salary differentials may not compensate for the difficulty of science studies at secondary and post-secondary levels. Obviously, modifying the reward structure for science graduates in the labour market, even in the public sector, is beyond the reach of many educational planners and policy-makers. Efforts could nevertheless be made to make information on opportunities in further and higher education, and in the labour market, more widely available at school level. Such information is badly needed in most countries and its absence makes choices depend on impressions and casual observations that may often be misleading.

Pupils who are performing well in mathematics and science at primary and lower secondary level are fewer in number than figures on participation indicate. Those who are genuinely committed to science may constitute an even smaller group. More than 40 per cent of lower secondary science teachers in Morocco indicated that only a small proportion of their pupils were actually interested in science; interviews with science teachers, students and counsellors in selected schools in Malaysia revealed that lack of interest in science and lack of confidence in performing well in these subjects were considered to be the main reasons for rejecting science and opting out (i.e. asking for transfer from science to other options).

A decrease in the proportion (but not in the absolute number) of the age group of secondary students taking science may, in some countries, be linked with fast progress in increasing participation in schooling of some groups – namely girls and rural children – who inflate overall student numbers in secondary education, but who may tend to be less likely to specialize in science than other students.

The situation of rural pupils deserves some comment. In both Malaysia and Morocco, the lowest proportions of students undergoing specialized science education at senior secondary level during the second

---

2. The test that students who apply for university or higher education courses in science and engineering have to take in addition to the academic aptitude test (PAA).

half of the 1980s were in the predominantly rural areas or provinces: in Malaysia, the overall transition rate into Form 4 Science averaged 24 per cent compared to 9 per cent for rural schools in 1991[3]; in Morocco in 1991/2, the predominantly rural Provinces (*Académies*) recorded proportions of senior secondary students entering the science stream that were clearly below the national average.[4]

The case study on Malaysia investigated the possible reasons for this. It revealed that science retains high status in rural schools (opting out of science after being selected is most prevalent in urban schools). The main problem is that rural students have great difficulties in meeting the entry requirements for selection into Form 4 science (and these requirements were rising in the late 1980s). Though facilities in rural schools, at least in Malaysia, are not greatly inferior to those in many urban areas, schoolteachers tend to be less qualified and staff turnover higher. Lack of genuine interest in science among children and the low levels of performance point to the need to make science curricula more attractive for everybody and to deploy special efforts to make mathematics and science education (from primary level on) more appropriate and accessible to all children, including those in the rural areas. Policies with this aim in mind will also have to take the special case of girls into account.

2.   The case of girls

Girls participate less and have lower achievement levels in science than boys in many countries. Policy-makers and planners must address this issue for various reasons (see Harding, 1992). Firstly, national development may be hampered if girls are discouraged from studying science and technology and their talents are lost to the labour market in these fields. Considerations of equity suggest that girls should not suffer discrimination, but should have equal opportunities of access to science qualifications and related employment. All citizens should have access to basic scientific and technological reasoning if they are to play a full part in development and not risk alienation by dependence on technologies which are not understood. Some authors (Harding, 1992; Keller, 1986) have furthermore argued that men and women bring different strengths

3.   Some rural students are enrolled in Special Science Schools, but the numbers are small.

4.   Forty-five per cent compared to 48 per cent.

and limitations to the generation of science knowledge and that the traditional domination of science by men may affect the conceptualization of existing problems, their study and their technological applications.

Participation in science by girls at upper secondary depends on their enrolment rate and standards of achievement at lower levels of the education system. Girls do not participate to the same extent as boys in lower secondary education (nor do they in primary education) in most of the sub-Saharan African countries, Papua New Guinea and Morocco (see *Table 3.1*). In France, Korea, Thailand, Jordan and Malaysia, girls and boys are almost equally represented at this level; in Botswana and Argentina, girls constitute a slightly higher proportion of general secondary students than boys (in these countries, boys are reported to be attracted by the alternatives to general secondary education – i.e. taking up a job or vocational studies).

The 'gender-gap' in participation increases at upper secondary level in many countries: girls form on average less than 20 per cent of the senior secondary students specialized in science in Burkina Faso and Papua New Guinea; at the other end of the scale, Thailand and Argentina record a majority of girls in this group (again, this may be because attractive alternatives to staying in school exist for boys). In France, more than half of the secondary students enrolled in the 'natural science' stream are girls (Caillods, Göttelmann-Duret, 1991). In Malaysia, girls are also evenly represented up to the end of upper secondary. However, in most countries girls are under-represented (with a percentage share of between 30 and 40 per cent) among secondary students specialized in science.

Where several possible science options and streams exist, the proportion of girls is particularly low among students specializing in physical sciences, mathematics and physics; the 'gender-gap' is smaller in biology, chemistry, in the natural sciences, and in integrated or general science. Generally, differentiation of options seems to contribute to broadening the gender-gap in science (see Harding, 1992). When there is a possibility of choice between 'hard' and 'soft' science subjects, girls tend to opt for the softer ones. In avoiding the more difficult and more selective 'hard science' options or streams (the C stream in the Francophone countries, for example) girls impede their own chances of access to the most prestigious training courses (e.g. engineering, informatics) at post-secondary level and their job and career prospects.

Table 3.1   Proportion of girls in the enrolment at different levels and in science streams (1990)

Countries	Percentage of girls in			Options or streams				
	Primary	Lower secondary	General upper secondary	Sciences	Mathe-matics	Physics	Biology	Chemistry
Burkina Faso	37.9	34.3	24.7		6.1(a)		15.7(b)	
Senegal	41.7		31.3		19.0(a)		27.2(b)	
Kenya	49.0		40.5	41.0(c)		33.0(d)		
Botswana	52.0	53.0	42.0	nd	nd	nd	nd	nd
Morocco	39.2	40.8	39.2	30.6	18.2		32.6	
Jordan	48.7	47.8	51.5	40.9				
Papua New Guinea	44.2	39.4	28.2	20.0	19.1	15.9	21.6	16.0
Thailand	48.6	48.7	53.2	55.0				
Malaysia	49.0	49.7	51.1	48.2/40.3(e)				
Korea	48.5	47.3	43.7	34.0/17.0(e)				
Argentina			57.7	64.7				
Mexico	48.4	48.6	44.5	nd				
France		49.4	56.3		36.9(a)		54.3(b)	

(a)   Maths and physics stream
(b)   Experimental science stream
(c)   2 science course.
(d)   3 science course.
(e)   Science School.
nd   no data
Source: Caillods and Göttelmann-Duret, 1991, and IIEP Country monographs.

The results of studies carried out within the IIEP project on the possible factors associated with girls' low participation in science courses converge with those of other research (see Adamu, 1992; Harding, 1992; Sharifah Maimunah and Lewin, 1993; Caillods, Göttelmann-Duret et al., 1997). They point to social and socio-cultural factors as determinants of girls' option choices. In her review of the literature, Harding (1992) argues that the social image and self-image of girls often remains dominated by their reproductive role in society. This keeps them away from secondary science studies which tend to lead to higher studies of longer duration (making marriage and child-bearing difficult) and to production-related jobs. Gender stereotyping is also often reinforced through textbooks and teaching materials which may not provide characters or situational examples with which girls can identify. The characteristics and attitudes of science teachers may also have an important impact on girls' interest in science. In many countries, females form a minority of secondary science teachers (see Ware, 1992b). The disparity is particularly strong among physics teachers and those teaching in senior secondary schools – irrespective of the science subject they teach. In Morocco, for example, the proportions of females among physical science and natural science teachers are 23 per cent and 45 per cent respectively at lower secondary, 17 per cent and 37 per cent at upper secondary level. The survey of science teachers carried out in Morocco also reveals that there are many science teachers – although these teachers are in a minority – who feel that science is more useful for boys than for girls; one-third at lower secondary and one quarter at upper secondary. In the case of Malaysia, the travel distance to upper secondary schools offering specialized science programmes proved more of a disincentive to girls than to boys in continuing to study science.

It is a widely held view that adolescent girls are 'naturally' disposed against science and are intellectually hampered in studying the subject. IIEP's evidence, and that from other studies, is strongly suggestive that this is not the case. Participation and achievement of girls in science differ widely between countries. This itself is evidence that such differences are culturally and institutionally determined rather than the result of any intrinsic differences between boys and girls. In France, differences in science achievement at the end of lower secondary can be explained much more by social origin and school-related factors than by gender (Duru-Bellat, 1994) . Even if there are differences in intellectual disposition between girls and boys, these would have to be very large to explain some of the disparities in performance that are observed. In any case, in some countries, girls perform as well as or better than boys. Once girls are enrolled in science programmes they often show similar or even higher pass rates in the secondary leaving examinations, e.g. in Morocco, Senegal, Jordan and Thailand[5].

Some interesting information on girls' achievement in science was collected through the latest international science achievement tests

---

5.　　Kenya, however, reported notable disparities between the pass rates of boys and girls in all science subjects.

conducted by the IEA. Three groups – 10-year-old pupils, 14-year-old pupils and students at the end of upper secondary education specialized in science – were studied. The test results present a very mixed picture (see Postlethwaite and Wiley, 1992): sex differences in science at the 10-year-old and 14-year-old level vary (even more so at upper secondary level). For many of the countries included in the IEA survey these differences are not the same at the three levels, or between the three main sciences. Sex differences in science achievement tend to be low at the ten-year-old level (in almost all cases, lower than among the 14-year-old pupils). Among senior secondary students specialized in science, sex differences are also relatively small in some countries, especially in biology. In one country (Israel), girls do even better than boys in certain subjects at this level. Differences in science achievement between schools are far greater in most developing countries than those between girls and boys.

A detailed picture of girls' achievement in science – based on appropriate performance indicators applied at different levels to all the science subjects – is lacking in the countries studied by IIEP. This is needed if disparities are conspicuous and this is considered problematic for the reasons discussed above. Where differential participation is a cause for concern, the reasons need investigation. They are likely to include consideration of the conditions of science teaching and learning and the socio-cultural environment in which girls are educated. The psychometric analysis of performance data on science examination questions can give useful insights. This type of analysis was conducted in Malaysia (on student samples drawn from the 1990 secondary school leaving examination in Integrated Science). It showed that all 10 questions on which there was the greatest gender difference in performance to the advantage of boys were physics-based; in physics, girls scored relatively poorly on questions relating to electricity. About 70 per cent of the questions where girls perform best are related to biology; this is convergent with girls' relatively good levels of performance in biology that the IEA study on science achievement observed in a large number of countries (among upper secondary students specialized in science). One possible conclusion to be drawn from this is that differences in science performance between girls and boys would diminish if more attention was given to the teaching of physics, and to electrical topics in particular and/or if the proportion of electrical questions in the examination considered was reduced (Sharifah Maimunah and Lewin, 1993). The latter begs the question of whether topics on which there are strong differences in performance should be reduced in importance or presented in more gender-neutral ways.

Some researchers argue that achievement differences between boys and girls (and more generally between different student groups) are not primarily subject-related, but are to a large extent due to differences in the construction of knowledge by these groups which are not taken into account in science teaching and testing. Girls, they argue, may typically see greater complexity in a scientific problem than do boys and try to weigh up rather than hierarchically order different aspects; they may be also less ready to ignore context. It may be presumed that girls' interest and achievement in science could be enhanced if teaching practice and materials recognized possible differences in cognitive style: such hypotheses have still to be validated empirically.

A few concluding remarks

(i)   Numbers studying science at upper secondary can be increased by directing more students to science-based options. This may increase the flow, but without additional measures to ensure quality standards may fall.

(ii)   Participation may be influenced by perceptions of subject difficulty and future opportunities. Curriculum development can be undertaken to increase the attractiveness of science, make difficult topics more accessible, enrich topics with examples relevant to students' lives, and introduce material related to possible science-based occupations.

(iii) Girls' participation may be increased by curricula that recognize their interests and cognitive styles; discourage girls from dropping science in favour of other subjects, thereby limiting future career opportunities; publicize role models of female science teachers and scientists; and enrich the presentation of topics that girls find especially difficult.

Section 3: Selected curriculum issues

Science curriculum development in developing countries has already a long history. Systematic attempts at curriculum reform in most countries date back to the 1960s and early 1970s, when many curriculum centres were established. Several recent trends have been identified (Lewin, 1992) which include:

- greater attention to 'science for all' programmes alongside continued emphasis on curricula designed for those who will continue to study science;

- widespread development of primary science curricula;
- more integrated/combined/co-ordinated/modular approaches, especially at lower secondary level, which draw content from the traditional science disciplines and stress scientific skills and cognitive processes above content;
- attempts to apply the results of educational research, especially on cognition, to curriculum design and special interventions;
- a new emphasis on the relationship between science and society and environmental issues in science curricula;
- technology as a new focus of interest in curriculum development, with an emphasis on skills needed to solve real-life problems;
- broader definitions of science education that absorb health education, nutrition, earth sciences, etc.;
- renewed concern for the quality and character of science teacher training.

Aspects of many of these developments are discussed in this study. Here, observations are made on some which have direct implications for the organization of science education and for the type of resources required: integration of content and inclusion of environmental education, the multiplication of options, and the introduction of technological concerns. Another important curriculum issue, related to the role of practical activity, will be dealt with separately in the following section.

1. Integration and relevance of content

Real-world problems that relate to development often do not have their basis in only one science discipline. It is also the case that at lower secondary level, the commonalities in aims and objectives between the science subjects are far greater than the differences, especially where these relate to cognitive skills rather than content. These considerations have accelerated the movement towards integrating lower secondary courses in which science is taught as a whole, without divisions into physics, chemistry and biology. This is now very widespread at the lower secondary level, with only a minority of countries continuing to adhere to separate subject teaching. At upper secondary level, integration is less common. The exception is for those who do not specialize who frequently experience general science courses. These typically group subject-based matter and often suffer from an inferior status to separate subject courses. It may be difficult to attract and retain adequately qualified science teachers to teach these curricula. This is problematic if levels of achievement are low[6] and if it is the last science instruction received by many before entering the labour market. This may be related to the lack of relevance of the content and/or to bad teaching practices.

There are several varieties of integrated courses – the integration may be based around concepts, or topics or may be little more than the mixing of traditional themes drawn from different science disciplines. Thus, the prominence given to the most obvious benefits of integrating – focus on science-reasoning skills which are promoted across a range of learning content – may vary. Several integrated curricula are well established and apparently successful. In the case of Botswana, the integrated science course introduces some applications of science and tackles multidisciplinary themes and topics. Kann and Nganunu (1991) report that although a certain number of implementation difficulties still need to be addressed, these courses have enhanced pupils' interest in science and attracted more young people towards science-related programmes and careers.

Environmental education, which has been promoted in numerous countries in the wake of the United Nations Conference on the Environment and Development, held in Rio de Janeiro in 1992, is a good example of some of the difficulties faced in the teaching of integrated and interdisciplinary content. Environmental aspects of science are reported to be part of science curricula (especially those at lower secondary level) in all countries included in the IIEP survey. This generally means that some topics chosen for study have an environmental dimension. The average rating of environmental topics in lower and upper secondary curricula in the countries covered by the IEA study of science is however moderate (see Rosier and Keeves, 1991), suggesting that this area is often added on to existing curricula as an afterthought rather than being assimilated as a coherent theme.

Deleage and Souchon (1993) suggest that the implementation of environmental education faces a variety of difficulties. The 'interdisciplinary' and application-oriented nature of many environmental curricula may sit uneasily with teachers' attachment to disciplinary based approaches and unfamiliarity with the non-scientific aspects of environmental education; most teacher training has hardly recognized the growth of work in this area; and the additional load imposed by new material on timetabling and teaching time has met with some resistance

---

6. In Malaysia, for example, well over 70 per cent of upper secondary students take this kind of course, and most reach very modest levels of achievement (Sharifah Maimunah; Lewin, 1993).

where it is not accompanied by reductions in other content. Examinations have also been slow to change to reflect environmental concerns.

Five major conditions for success in the introduction of environmental science curricula, content and concepts are suggested by Deleage and Souchon (1993), in their paper:

(i)   a clear political will to promote environmental education;
(ii) public opinion supporting this policy;
(iii) progressive 'infusion' of environmental education into existing subject matter;
(iv) consequent incremental changes in teacher training (with priority given to the training of teacher trainers), school organization and timetabling;
(v)   development of appropriate materials (not only textbooks and teacher guides but also specialized documentation and newsletters suggesting practical experiences).

2.   Multiplication of options

As shown in *Chapter II*, an increasing number of countries are introducing options as an alternative to streams. The basic objective is to respond to a diversity of pupils' interests, allow them to test their potentiality and at the same time enable them to create their own individualized pathway into further education or employment. The introduction of open electives in a curriculum allows schools themselves to exercise their originality and to adapt the programme they offer to the specificities of the local environment. In some countries, the multiplication of options is the result of the choices offered at the examination. As a result, some systems offer a very wide variety of electives. In many English-speaking African countries, such as Botswana or Kenya, the following subjects could have been offered in a school: physics, biology, chemistry, integrated science, agricultural science, physical science (physics chemistry), biological science, biology-physics, human and social biology, combined science, additional combined science. The disadvantages of such a system in a resource-poor country, or even in a middle-income country, are clear.

First, problems of teacher deployment may become insurmountable: at best, schools offer electives on the basis of the qualifications of their teachers, in which case, it is not the pupil who chooses according to his or her specific interest, nor are the subjects chosen necessarily the best ones to be taught in the school's specific environment; at worst, subjects are taught by teachers who have no qualification to teach them.

Second, inequalities between small (generally rural) schools and large ones (generally urban) become more marked: given their enrolment, small schools cannot afford to offer more than one or two electives without lowering student numbers in the different groups to unrealistic levels; large schools, on the other hand, are in a position to offer to their pupils a much larger range of choice. Also prestigious and expensive options (such as computer science, for example) are only offered in a few schools, generally those attended by the most privileged pupils; this may increase the inequalities of access.

Third, students' timetables become overloaded with electives: students may tend to accumulate such electives in the hope of obtaining more credits or points in examinations.

Fourth, students playing the option system try to maximize their chances of success and tend to select subjects which are known for their much higher pass rates. This may be to the detriment of science electives. In Malaysia, for example, students wanting to maximize their chances of passing the examination, may select principle of accounts rather than, say, physics, since the pass rates in the former subject are higher than in the latter.

Fifth, other disadvantages of option systems are that where assessment rubrics have many permutations of options, it is difficult to assess how many students are studying how much science and in what depth. Extensive choice of options or electives makes the planning of provision difficult and usually increases unit costs, since group sizes are smaller.

Finally, comparability in selection procedures between candidates is problematic and monitoring of core standards becomes technically very complex.

The acquisition of scientific literacy of a minimal kind would seem to depend on all pupils studying science at least to the end of lower secondary, and preferably until the end of upper secondary. As the rubric of elective subjects and their weighting is likely to determine the numbers of students studying science and the length of time they persist, there may very well be a contradiction between an extensive choice of options and the objective of science literacy for all.

The following questions seem therefore relevant for planners and policy-makers. Why are so many options offered in the curriculum or in the examination? Are the enrolment ratios so high that it is indeed necessary to respond to a diversity of students' interests? How much is it going to cost? Are the resources adequate to allow such a system to be sustained without greatly increasing the disparities between schools and regions?

It is often unclear what the educational and human development rationales are for a wide choice of science options.

3. Technology education

School science has often been criticized for being too abstract, too academic, having a weak relationship with the outside world that pupils experience, and for failing to lead to personal and collective empowerment with respect to urgent practical problems (Layton, 1994). 'Learning why' or 'knowing that', a traditional concern of science education is not the same as 'knowing how', nor does it necessarily lead to it. Since the 1970s and 1980s, industrialized and developing countries have shared a common concern, which is to develop practical knowledge and capabilities within general education, something called 'operacy' by analogy to literacy and numeracy (Knamiller cited by Layton, 1994).

An international survey carried out by UNESCO in the mid-1980s (UNESCO, 1985) indicated that some form of 'technology education' was part of lower secondary curricula in all regions of the world. In Latin America and Africa, the percentage of time devoted to 'technology' appeared to be slightly higher than in Asia, Europe and the Arab Region (14 per cent; 10 per cent; against 9 per cent; 8 per cent; 6 per cent). The objectives, approach and content of what is called technology education, however, vary widely in the different countries: from craft courses which aim at teaching a range of practical skills, emphasizing the development of simple pre-vocational skills and including some proficiency in the use of tools (e.g. Kenya), to more general courses emphasizing design, understanding of general technological concepts and the solving of technological problems (e.g. England and Wales), up to renewed, updated, technical courses at upper or post-secondary level, introducing flexible specialization instead of job-specific training, and helping students to understand the concepts that form the basis for the development of products (e.g. France, Colombia).

The concerns behind the introduction of technology education are numerous.

Some are *economic*: the need to train a more flexible, adaptable workforce. Martinand (1994) notes that in many developed countries traditional 'technology' or 'technical education' courses aimed at developing craft skills have been increasingly criticized, and their detractors have urged that they become more general, basically for two reasons. First, over the last two decades rapid technological change has led to more and more job mobility and less need for job-specific skill requirements; the technical skills needed for almost any employment are changing faster than ever before and it is becoming more difficult to predict which specific skills may be needed in the future. Flexible specialization as an outcome of technology training has become more attractive than mono-skill certification. Other concerns are more *pedagogical*: it is felt that many pupils find meaning and relevance in their studies only to the extent that they are related to the world they know. Practical activities would enhance their interest and understanding of science, for example. Other concerns are more *social*, related to the need for raising the awareness of and control over the technological innovations that shape peoples' lives (Layton, 1993; Ware, 1992a). Attempts have thus been made to introduce a critical approach to the study of technology and its applications, and to design and implement Science-Technology-Society (STS) courses – placing technology in its social, political, environmental and moral context.

In most developing countries, however, the emphasis is on closing the technological gap, facilitating transition from school to work, and enhancing self-reliance. The main objectives assigned to technology courses seem generally to promote children's interest in practical work and to develop the (pre-vocational) skills and attitudes that are required for work in agriculture, industry or other production sectors. The case study on technology education in Kenya provides some insight into the objectives of such programmes, but also into the difficulty of implementing them in resource-poor countries (Obura, 1992).

The Kenyan Government has tried to give increasing attention to technology education and training measures aimed at enhancing adaptability and improving the life chances of school leavers. The Gachati Report that led eventually to the education reform of the mid-1980s placed particular emphasis on the development of pre-vocational craft-oriented skills, including those relevant to small-scale businesses. The policy option adopted was to provide courses in agriculture, agro-crafts, commercial activities and services in the primary curriculum; at secondary level, a pre-vocational orientation was to be continued and a large variety of options, in particular 'industrial education', were to be offered in order to make school leavers more attractive to employers. Science-oriented problem-solving methods were to be stressed.

Under the current curriculum, pupils in upper primary (standards 6 to 8) spend considerable time on art, craft, agriculture, home science and business studies. At secondary education level, students in Forms I and II, until 1993, had to take agriculture, and another technical subject to be chosen among a large variety of technical options (woodwork, metalwork, building construction, electricity, power mechanics, drawing and design, home science) to which a third 'practical skills subject' –

either business studies, music or foreign languages – had to be added. At least one practical subject was compulsory during the last two years of the secondary cycle. When it came to implementation, this programme had to face enormous difficulties due to the large number of options to be offered, the lack of trained teachers, the dearth of equipment and raw material, and an apparent lack of support from parents and teachers. The problems were most acute in the less privileged schools where the largest numbers of secondary entrants were enrolled and who were thought most likely to benefit from technology education and to pursue middle-level job opportunities. Paradoxically, pupils who selected three separate science subjects in more privileged schools could – and in many cases actually did – avoid technology-related subjects during the last two years. A new curriculum was introduced in 1993 which slightly increased the time for technology subjects while reducing the range of subject choice in secondary schooling (agriculture is no longer compulsory but will in reality remain among the very few technology subjects that are actually offered in the less well-equipped schools). The curriculum reorientation attempts to respond to some of the implementation difficulties that the former curriculum with its variety of options faced (Obura, 1992; Wanjala Kerre, 1994).

Other experiences with the introduction of technology education exist in Africa which have a slightly different focus. Many of these programmes face similar difficulties in implementation (Robson, 1992).

Central issues to be addressed by planners and policy-makers with respect to technology education are:

(i) Should technology be taught as a separate subject? To what extent can science curricula accommodate the main aims of technology education at any particular level in a cost-effective way?

(ii) Should technology courses be provided as part of the compulsory curriculum at lower and upper secondary levels? Should they be organized at a post-compulsory stage in parallel with academic programmes?

On the *first question,* the arguments are clear in principle. Treating technology as a separate subject has the advantage of treating "the nature of technological knowledge as a unique and irreducible mode" (Layton, 1993). It is claimed also that technology will be fully recognized only if it is allowed to develop its own curriculum space. On the other hand, treating it separately increases the burden on curriculum time if all students are to follow some technology-related courses. Space has to be found in what are often already overcrowded timetables. Depending on the approach, more or less sophisticated facilities and equipment will be required in addition to those needed for the teaching of science. If these are conventionally conceived their costs may be high. Maintenance of equipment has also to be funded, as well as the costs of consumable materials. High-income industrialized countries and middle-income countries (such as Malaysia, and Latin-American countries), which enrol large proportions of their relevant age group and which have sufficient resources, are increasingly embarking on such programmes at the lower and upper levels of secondary education, either in main-stream general education or through a separate track. However, most countries do not have trained cadres of technology teachers outside those involved in traditional subject areas where the emphasis has often been on craft skills related to single occupations (carpentry, metalwork, typing, etc.). It may be difficult to involve these (often under-qualified) staff into higher-level technology teaching, where flexible skill application and development and design capabilities are stressed. In the short term, teacher supply constraints may limit the growth of technology enrolments at school level.

In low-income countries, the problems of increasing the amount of technological teaching may appear insurmountable, at least in the short term. Where qualified and experienced teachers are lacking, where these are obliged to accumulate additional teaching periods or seek a second job to make a living, where in-service training is scarce, and equipment missing or insufficient for the teaching of existing science subjects, it may be unrealistic to add a new subject to the curriculum. Promoting science education as a separate subject might even contribute to making science education more academic and separated from the day-to-day life of pupils.

If these conditions exist, then technologizing the science curriculum may well be a much more realistic and cost-effective alternative to introducing technology as a separate subject, especially at lower secondary level. It has the advantage that a basic grounding in science is usually a prerequisite for purposeful application of technological skill based on analysis rather than intuition. It simplifies the logistics of providing some technological insights for students and it offers the prospect of providing technological awareness to those who subsequently specialize in academic science. It may also succeed in offering technological learning experiences to more girls for longer than will be the case where technology is taught as a separate optional subject. Martinand (1994) concludes that plans to introduce technology education into general secondary education should be modest. Until the end of lower secondary, applied science curricula can be an alternative in which opportunities can be given for some project work involving students in practical problems in order to develop their practical capability. The availability of appropriate facilities, materials and human resources are prerequisites of success in curriculum innovation and these could be more easily provided if technology is incorporated within science.

On the *second* question, the answer is also clear as far as lower secondary education is concerned: the optimal solution seems to be to integrate technology into the curriculum of general education. But it is not quite as clear as far as upper secondary education is concerned. Different practices have evolved in different countries, depending on their tradition and culture, their level of economic development and on the requirements of their labour markets. Most continental European countries have retained different tracks, while the Anglo-American tradition has tended to introduce related options in their general school curriculum. Keeping technology education in a special separate track allows for a more product-oriented concept of technology, using modern types of technological equipment. According to Martinand (1994), at higher levels, the importance of specialization in areas of technology may weigh in favour of provision separated from mainstream science curricula. The determination and content of technology courses has to reflect overall development strategies and realistic appraisals of the labour market for those who qualify.

The question of providing practical, design and technological activities has acquired a new dimension as countries' chances of developing and integrating into the world market are increasingly seen as a process intimately bound to international competitiveness. Successful economic transformation is likely to be accompanied by the introduction of new methods of organizing production, in industry, and in all probability in other sectors too. These methods stress flexible specialization and democratization of the workplace in ways which capitalize on the existence of an 'intelligent workforce', quality control as the responsibility of all, and commitment to continuous improvements in process. If this is true, it carries implications for technology and science education, as Lewin (1995) argues for South Africa. Some examples illustrate the point:

> Intrinsic standards of quality control can be encouraged through carefully designed practical work. However, it remains a rarity for students to discuss positively the quality of their work or that of their peers, or to have the time and opportunity to reflect critically on their own performance. Judgements of quality are generally externally defined against criteria which may not be very transparent. Quality control in practical science is often

synonymous with getting expected answers rather than anything more richly reflective. Methods of assessing practical science and technology have therefore to be revised and improved.

And in school science it is rare for experiments to be repeated with the intention of improving design, the precision with which measurements are taken, or to stress a system to explore the weakest links in a chain of analytical procedures. It is also unusual for students at any level in schools to contribute to the design and development of experiments since time is rarely sufficient. These kinds of skills are more likely to be developed in a high-quality technological school.

Moreover, team work is central to the concept of Total Quality Control. Flexible specialization, allowing both rotation of jobs between individuals and the performance of several processes using different skills by the same individual, implies collaborative working practices. Often, especially in the higher grades and near to public examinations, little such collaboration is evident. Co-operative learning strategies are replaced by competitive ones. Technology may be more likely to encourage collaborative working through small groups work on meaningful problems and projects (technical or industrial projects).

All of the above may be interpreted to favour introducing separate technology courses or tracks at upper secondary level, where special attention will be paid to questions of design, measurement and technological problem solving. It also stresses the need to pay more attention to the way practical activities are conducted in science, whether or not they are combined with technology.

Section 4: The role of practical activity in science education

Problems in planning science education cannot be approached separately from some consideration of the role of practical activity in learning and teaching. If the costs of teaching science are relatively high when compared to other subjects, it is primarily because of the investment that takes place in providing and maintaining practical facilities and the implications these carry for the employment of ancillary staff. In some systems it is also because of the smaller group size associated with practical subjects. The high costs are therefore a direct product of curricula and organizational decisions on the role and type of practical work in science education.

The main purposes usually given to justify practical work in science education can be encapsulated in many ways. Haddad and Za'rour (1986) identify four assumptions related to practical activity:

(a) it fulfils the stated objectives of science teaching, especially those related to inquiry and discovery;
(b) it is necessary because science is essentially experimental;
(c) it is justified on psychological and pedagogical grounds;
(d) it has positive effects on educational outcomes that can be empirically verified.

They note that the *first* often assumes an unrealistically wide exposure to practical work which cannot be achieved through alternative (and less costly) teaching methods. It sometimes appears that students are expected to acquire most of the skills of a professional scientist through an hour or two a week of practical work. The IIEP study corroborates the observation that provision of practical facilities is no guarantee that relevant objectives will be achieved or that practical work will actually be undertaken with an appropriate frequency or quality.

The *second* risks the devaluation of the non-experimental aspects of science – conceptualizing, modelling, theoretical analysis – in favour of what some have labelled "privileging the practical" (King et al., 1989). Practical work may be reified to the extent that it is held that no science lesson is complete without a practical activity, no matter how trivial or conceptually suspect. Curricula are planned with a practical activity for almost every lesson, though it is unlikely that all will be conducted in ways which link theorising with empirical experience.

The *third* assumption Haddad and Za'rour partly dismiss on the grounds that "adolescents who have moved well into the stage of formal operations should be able to think abstractly without the need of referral to objects to aid in conceptualizing or abstracting". This assumes most adolescents studying science have reached the Piagetian formal operations stage, which is false. The best data available in the United Kingdom suggest that only a minority of secondary school students consistently practice formal operational thinking (Shayer and Adey, 1981). The proportion is unlikely to be much greater elsewhere. However, Haddad and Za'rour are correct to note that Ausubel's 'meaningful verbal learning' (i.e. coherent bodies of propositions) (Ausubel, 1968) does not depend on practical work (see below).

Though practical activity may have motivational benefits, there is no convincing empirical evidence that this is always the case. Thus it is not clear that, for example, less able science students have a preference for experimentation or that it sustains their interest in science (Lewin, 1992). Head (1985) argues that typical school science practical activity actually becomes more formal and less exploratory just as adolescents become less conformist and more aware of their own individuality and capability of taking risks. It therefore becomes less attractive to those not already committed to science specialization.

The *fourth* is not at all easy to demonstrate. As noted earlier, higher levels of practical activity and provision are generally not associated with higher levels of achievement on conventional assessments. Though this empirical finding may be partly spurious (as outcomes related to practical work may not be assessed, see below), the absence of a simple association is worrying, considering the investment of time and money given to practical work.

The Cognitive Acceleration in Science Education (CASE) studies from the United Kingdom provide some support for the importance of practical activity, albeit that associated with simple and cheap experimentation and 'thought experiments' at the lower secondary level. It appears that the cognitive conflict created in students' minds when observation contradicts prediction may be an advantage in developing cognitive strategies for problem solving. It may also be a basis for the 'bridging' of concepts to new contexts and the spreading of general thinking skills to other subjects. CASE data show significant gains in subsequent science examinations for those experiencing a relatively small number of intervention lessons in the context of normal (to the United Kingdom) amounts of practical science, using a control group comparison design. They also indicate the possible transfer of thinking skills to other subjects (Adey and Shayer, 1994).

There is extensive literature on practical work and its effects – too vast to review here. A recent contribution draws attention to the importance of distinctions between different types of knowledge which may imply different conditions of learning. Ausubel's (1968) 'meaningful verbal learning' has been contrasted with Gagne's (1965) intellectual skills, which involve tightly defined algorithmic procedures for carrying out classes of tasks (White, 1991). The question is whether either type benefits from associated practical activity, as is often assumed. White's view draws attention to other types of knowledge – images, episodes, strings, motor skills, and cognitive strategies – that may act to mediate the acquisition of intellectual skills. Of these, he argues that episodes may be especially important when considering the role of practical work. Episodes are recollections of events that help locate other types of learning and assist in their interpretation. The apt analogy is that of a sunset that can be described in detail but when it is observed it takes on a different set of meanings which become part of an understanding of the construct. Practical work therefore provides opportunities for episodes to be incorporated into the mental map of science constructs students acquire – but only if they are located within appropriate conceptual frameworks.

Other authors also stress the importance of direct experience and the 'personal response' to science that may result from practical work. Thus Head (1985) suggests that carefully arranged practical work can create the conditions for meaningful rather than rote learning and shift the balance from the reception of information to interaction and manipulation of ideas (Novak, cited in Avalos, 1995). This fits in with Bruner's ideas of 'inactive' learning acquired through the bodily experience which is characteristic of young children's learning. Like many others, Head notes that poorly conceived practical work is unlikely to result in meaningful learning if students are unaware of its purpose and have no cognitive map to which the learning experiences relate coherently.

Allsop offers a somewhat different but overlapping list of purposes for practical activity to Haddad and Za'rour. These are stimulating interest and enjoyment, learning experimental skills and techniques, teaching the processes of science, and supporting theoretical learning (Allsop, 1991). However, he goes on to observe that: "the prevailing paradigm of inquiry– orientated practical science may overlap uncomfortably onto the teaching and learning modes associated with instruction in the ... pre-industrial societies of many low-income countries" (ibid). His thesis is that often science teachers have attended relatively well-equipped secondary schools and have expectations of teaching science in modern fully-equipped laboratories. Those who train them have rarely spent a long time in typical resource-poor schools and have limited understanding and sympathy with the problems of teaching science without laboratory facilities. This, coupled with an attachment to evaluation through summative written tests, militates against effective practical work in under-equipped schools.

Knamiller (1988) in his work in Malawi has shown that investigations can be devised which are based on local science and technology and which are inexpensive and accessible. The problem is often one of generating enthusiasm and confidence in such approaches in competition with the prevailing orthodoxy that science has to be taught using laboratories in secondary schools. Swift (1983) working in Kenya, has developed a secondary physics programme where all the experimental work has been chosen to relate to a familiar context for rural children and to be undertaken using locally available material. The Science Education

Programme for Africa (SEPA) also developed a range of models for teaching science in line with these ideas.

The message from this short review of the role of practical work is that traditional laboratory-based school science may be 'over-engineered' if value is placed on the thinking skills that it can promote. Despite the wide range of initiatives that have occurred to simplify practical activity, the data assembled in this study suggest that in many countries the teaching of science in a conventional laboratory-based environment remains the normal expectation. This is notwithstanding widespread inability to finance the costs for all except a small minority of students. Where laboratory costs are many multiples of the cost of ordinary classrooms, running costs cannot be sustained, and where laboratories are often used as ordinary classrooms by science teachers the purpose of high-cost laboratory-based provision must be questioned.

Ultimately, the problem is a learning and teaching problem rooted in curriculum expectations and patterns of assessment rather than a more physical one. Too often, practical activity is confused with laboratory work, and the image of laboratory-based science shifts the focus away from the thinking strategies and manipulative skills that are probably the most important goals of practical activity. Active engagement with problems in the physical world is part of everyday experience and most, if not all, worthwhile thinking skills associated with secondary science can be taught without expensive equipment. Many lower-cost alternatives are available, and demonstrations, videos and other simulations may be at least as effective but cheaper than individual or group practical work in promoting conceptual development (Lunetta and Hofstein, 1990). The changing nature of competitive economic activity carries with it implications for the character of practical activity. However, these are qualitative in nature and need not require high-cost provision of practical facilities.

Certain questions confront the planner faced with demands for high-cost laboratory provision, which need a considered analysis before agreeing to finance provision:

- what are the curriculum objectives that laboratory work can contribute to and which could not be attained if lower-cost facilities are provided?
- are fewer, higher quality 'minds on' practical activities preferable to large numbers of poorly delivered practical activities that are often 'minds off'?
- are the running costs of planned practical facilities likely to be sustainable?

- what implications for training and teacher support stem from policy on practical facilities?
- how and why is practical activity to be assessed?

Section 5: Assessment issues

There are a number of issues in assessment that have particular significance for the planning of science at secondary level. The most important of these are concerned with:

- the significance of assessment in science for progression through the school system;
- the relationships between the curriculum and patterns of assessment;
- the problems of assessing practical work;
- the ways in which assessment information from national examinations and from purpose-built monitoring tests can be used to plan interventions to improve quality and performance.

These are discussed below.

1. Science assessment and progression

Key issues in relation to assessment and progression are the extent to which assessment instruments reliably select students for science on the basis of valid tasks. In particular: (i) which subjects and what weighting should be used for selection decisions; (ii) to what extent are the tests administered assessing the whole range of skills and knowledge necessary for further science studies?

The first problem concerns the weight given to the assessment of science and mathematics capability in selection tests. As seen earlier in *Chapter II*, the majority of countries assign relatively high weighting to mathematics and science in selection examinations for upper secondary education. Very often mathematics accounts for more than science. In other words, access to secondary courses usually requires more evidence of achievement in mathematics than in science. Once students have obtained admission to upper secondary, their access to different science streams or options may depend on different criteria, according to the specific case considered. In countries following an Anglophone or American pattern, allocation is the result of a mixture of preference and satisfactory minimum performance in lower secondary mathematics and science. In Francophone systems, access to selective science-based streams generally requires high overall performance, although mathematics and science subjects are given extra weighting.

For entry into post-secondary science studies, the key question is whether selection is best based on students' achievement in the science subjects or on a much broader range of subjects. In Francophone countries, selection into science courses will weight science and mathematics around 50 per cent in the scientific baccalauréats. This means that students can enrol in post-secondary science courses despite mediocre achievements in science (and/or mathematics). This may have an impact on progression rates and internal efficiency. The IIEP study on the destination of school leavers in Morocco reveals that a high proportion of secondary school graduates enter higher education programmes despite mathematics and science scores below average; at the same time, there are extremely high failure rates among secondary science graduates at post-secondary level (Caillods, Göttelmann-Duret et al., 1997). This correlation could not be explored in depth; it is also not the only intervening factor. However, it suggests that the selection process may be influential in creating conditions for subsequent repetition and drop out. Anglophone countries sometimes maintain a pre-university level (A Level) at the end of upper secondary which is highly specialized. In this case, students will often only take mathematics and science subjects for entry to university. Typically, failure and drop-out rates in Anglophone tertiary science institutions are lower, perhaps as a result of greater selection.

The form and content of assessment on which selection into science courses is based is another critical issue. Evidence from a variety of studies (reviewed in Lewin, 1992) suggests that most public examination systems fail to cover the full range of goals to which science teaching and learning at the different levels is directed. There are several reasons for this. Firstly, the fact that assessment at upper secondary level is largely undertaken through written tests places boundaries around what can be assessed; certain skills which are important in science-based activities, such as communication or co-operative working skills, can only be tested to a very limited extent; multiple-choice instruments tend to be particularly restrictive in this respect. In most countries, students' science scores at the end of lower or upper secondary depend to a very minimal extent on their achievement in experimental and other practical work.

Research on public secondary examinations which was carried out in the early 1980s in several English-speaking African countries indicates that common examining practice in science is dominated by recall items (ILO, 1981). This continues to be the case at least at the primary/secondary divide (Bude and Lewin, 1996). The analysis of science examination papers administered at baccalauréat levels in Morocco confirmed the persistent importance of memorization and reproduction of factual knowledge in science testing, although the proportion of recall questions varied from one Académie/region to another – and even more within one Académie/region from one year to another. Analysis of Malaysian examination items also indicated a preponderance of recall items in lower secondary examinations and a substantial number at upper secondary.

The power of discrimination of selection examinations, and their quality, needs close consideration. There may be some items which are completed relatively well by rural students, despite the fact that their performance on other items is inferior to that of urban students. Does this indicate that differences in performance are arising partly from the specific items chosen and the method of presentation or are the differences real? The same reasoning can be applied to observed differences in performance between boys and girls. In Malaysia, most of the difference in performance between high-scoring boys and girls arises from less than 10 per cent of the curriculum as measured by the content of examination items. Planned intervention to reduce disparities in performance and successful selection into science needs to pay particular attention to possible 'biases' in examination items. The point here is not to delete those items that prove to discriminate against certain pupil groups, providing this is evidence of underlying learning difficulties and not item presentation since the content and skills may be relevant things to test. The objective should be to improve teaching/learning processes in areas where differences appear greatest if the goal is to reduce disparities in performance.

In principle, continuous assessment – which distributes teacher or school-based assessment over a period – should be more reliable and valid than brief external standardized examinations. It can cover a broad range of assessable science outcomes and provide a more comprehensive picture of students' ability to participate successfully in further science studies. A number of countries have replaced external (provincial or national) examinations either partly (e.g. in Papua New Guinea) or completely (e.g. in Morocco) by systems of a more school-based assessment, especially at the stage of transition from lower to upper secondary education. However, whether this is an appropriate alternative depends on the competence and willingness of classroom teachers. Continuous assessment may be less objective than external examinations. The conditions for effective implementation are often difficult to achieve and assume well-trained teachers, good record keeping, professional ethics, and adequate moderating procedures.

In Morocco, pupils' access to upper secondary education and their tracking depends on a mix of continuous and end-of-term school-based assessment. Yet 93 per cent of lower secondary science teachers indicated that they felt insufficiently prepared for the assessment tasks they have to undertake; almost 50 per cent said that they had received no training at all in the area of pupil assessment and considered the official instructions concerning pupils' assessment in science subjects as 'not clear at all' (Caillods, Göttelmann-Duret et al., 1997). One of the effects of this has been a large variation between schools and between teachers concerning the types of science skills actually tested at the end of lower secondary. Increased workloads for teachers and insufficient time set aside for assessment tasks are further obstacles which often impinge upon effective implementation of teacher or school-based assessment.

School-based examining moderated by national examinations has been tried out in Papua New Guinea. This can offer some of the benefits of continuous assessment (range, authentic assessments under normal conditions, relevance to the curriculum as taught), and of external assessment (objectivity, standardized comparisons of performance, curriculum monitoring) and may offer an interesting solution to the problem of reconciling the need to adapt assessment to specific local or school-specific conditions, on the one hand, and the need to ensure equitable and comparable standards between teachers and schools, on the other. For selection to be valid and reliable, this option is only worth considering if adequate infrastructure exists to support its introduction. It will take time to be implemented and requires minimum levels of competence in the design and application of assessment techniques by teachers. This issue will be discussed below.

2. Assessment and the curriculum

The forms that examinations and assessment take are widely recognized as determinants of educational practice. How secondary science is taught, which topics are emphasized, and which scientific knowledge and skills are developed are all likely to be strongly influenced by the nature of what is assessed. Extensive literature exists on the general effects of assessment on teaching and learning in developing countries (e.g. Dore, 1976; Oxenham, 1984; Kellaghan and Greaney, 1992) which illustrates the degree to which educational practice may be determined by the choices made in measuring outcomes.

In science subjects, experience with the inter-relationships between assessment and curriculum development has a mixed history. The science curriculum reform movements of the 1960s and 1970s stimulated the development of new curricula throughout the developing world that placed an emphasis on guided discovery and scientific problem-solving skills and reduced the stress on factual content (Lewin, 1992). In most countries, examination systems were slow to change to reflect these new emphases. It has been relatively unusual for a change in examination practice to be used to promote and reinforce desired changes in learning and teaching (Black, 1990). Often the main reasons for changes in examining structures lie in the need to reduce costs, maintain security and increase reliability, rather than because new curricula demand new styles of assessment.

The development of the first generation of new science curricula coincided with the widespread adoption of examinations using multiple-choice items which could be machine marked. The majority of education systems following the Anglo-American tradition adopted this style for assessment in science at the end of the primary cycle and for lower secondary; many depended on it as a major component of assessment at upper secondary. Multiple-choice items cannot easily be used to assess the full range of educational outcomes that are commonly associated with new science curricula. For example, though creative item writers can design items that assess some aspects of problem-solving skills and experimental design, there are clearly limits to the validity of this process. Moreover, where item writers are inexperienced and lack sufficient development time and training, there may be a tendency for assessment instruments to contain a predominance of items which assess recognition and recall, and relatively few which assess the acquisition of higher order skills. The evidence in the early 1980s suggested that in many Sub-Saharan African countries, science examinations remained heavily recall-orientated despite the curriculum development that had taken place (ILO, 1981). Though there were some exceptions, e.g. in Kenya, creative development of high-quality items to assess a broad range of educational outcomes was exceptional. A more recent review (Kellaghan and Greaney, 1992) indicates that there may have been some improvement but there is still frequently a mismatch between the emphases of assessment and those of new science curricula.

Two aspects of the curriculum/assessment relationship in science subjects stand out. First, the kinds of outcomes that science curricula often aspire to can only generally be assessed through a combination of written tasks under examination conditions, performance of practically based activities, and study and reporting activities that may extend over a relatively long time and involve collaborative work with others. Thus, many outcomes will not be assessed unless it is possible to introduce some elements of school-based examining. This is common in many

127

school systems in developed countries but relatively uncommon in developing countries. In Section 3.2, the heterogeneity of school-based science tests administered at the end of lower secondary were mentioned. A recent review (Pennycuick, 1990) also draws attention to the range of approaches used and some of the problems that have arisen. Prominent amongst these are the increased assessment burden on teachers, reduced reliability, more sophisticated moderation, and the prevention of collusion, cheating and favouritism. Furthermore, school-based assessment may favour children with richer home resources, especially where assessment includes extended project work since educated parents can provide more access to resources and contribute supportive advice.

The second aspect is concerned with opportunities to reshape the curriculum/assessment relationship through the introduction of curricula based on criteria of competency. Science curricula are often judged to be susceptible to this kind of approach because it is thought that competencies can be defined unambiguously. These developments were foreshadowed by more general work on mastery learning which has a long history. They are closely associated with the introduction of criterion-referenced assessment as an alternative to norm-referenced testing.

At first sight, competency-based approaches appear attractive and feasible. In principle, the invitation is to define performance outcomes that profile competency at a particular level by specifying statements of mastery within relevant domains. These then provide the criterion behaviours against which students' performance can be assessed. The curriculum, and certainly the assessment instruments, almost 'write themselves' since the structure and content of items is suggested by domain statements.

What seems a simple and logical approach to redefining the relationship between assessment and curricula turns out to have many difficulties in practice. These have been widely noted (Frith and Mackintosh, 1984; Black and Dockerell, 1984; Wolf, 1993; Lewin 1994). 'Domain' statements in science are not all simple statements which can define competency – being "able to calculate an appropriate value for a fuse using Ohm's law" seems not too difficult to convert into an assessment task that is reasonably unambiguous[7]. However, developing an understanding of energy conversion, or attributing meaning to

7. This is actually not as easy as it may appear. The conditions under which the task is performed need to be specified, the precision expected has to be defined, and what is appropriate demands knowledge of conventions.

relationships between living things within an ecosystem are far more difficult to define in terms of simple competency statements. Technically, whilst closed domains[8] can usually be specified with some precision, open domains[9], which involve higher levels of cognitive behaviour, are often difficult to define. One risk is that attempts to achieve increased precision, result in a very lengthy list of competencies, so long as to lose much practical application. Some research (Wolf, 1993) suggests that operational definitions of competency are often understood more readily from examples of typical assessment items than from domain statements that are logically supposed to precede the construction of such items. What appears a deductive approach to curriculum development (start with competency statements, derive learning activities, specify assessment tasks), in practice, may often become more inductive (where examples of the type of performance, accepted as evidence of mastery, lead to learning and teaching activities which can be used to define statements of attainment).

Competency-based approaches to curriculum development may also have difficulty in defining outcomes that are both sufficiently challenging to test the most able students of a given age, and not too difficult so as to discourage completely the less able. Competency may be defined in terms of successful completion of a related set of particular tasks, such that each one has to be performed to an acceptable standard. If this is so, either these tasks will have to be set at a level where they will be trivial for many students so that most are judged competent (with the attendant risk that the minimum level of achievement becomes the maximum) or the tasks may be too difficult for many if competence criteria are derived from what ideally should be mastered.

This kind of problem will be exacerbated if minimum learning criteria are defined in ways that create an expectation of a single level of performance for a grade group of children. Achievement within a cohort will be distributed across a range. No single definition of minimum competency criteria is likely to be satisfactory since it will not reflect this range but divide it into those with competence and those without, failing to recognize questions of degree and progress towards levels of understanding. The system currently adopted in Mauritius (based on essential and desirable learning competencies) has this problem. Other systems (e.g. the United Kingdom National Curriculum) do try to

8.   Simply those with a single relatively unambiguous performance criteria.

9.   Those domains where many outcomes could indicate mastery.

accommodate this difficulty by specifying a band of levels of achievement that is considered normal for students of a particular age.

Analysis of the relationship between curriculum and assessment leads to several conclusions. *First*, the development of more school-based assessment is attractive because it offers opportunities to test a full range of outcomes from science programmes. In most cases it is desirable to couple the development of this with periodic standardized assessments which can be used to monitor standards and moderate internal assessment scores – as was the case in Papua New Guinea with the use of the Mid-Year Rating Examination to moderate school-based assessment (Lewin, 1992). Adoption of school-based assessment will be limited by the development of adequate educational infrastructure, the extent of training of science teachers in assessment techniques, and public acceptability of school-based assessment, especially if it has selection consequences. In those countries where large numbers of science teachers are untrained and communications are poor it may be that the best advice is to retain external examinations and concentrate on improving their quality, reliability and validity. Monitoring tests are also attractive to review periodically standards of achievement across schools, districts and regions.

Second, movement towards more criterion-referenced, competency-based assessment is desirable since it focuses attention on intended outcomes and actual levels of achievement. It may become technically complex and cannot be mechanistically implemented. Nevertheless, it does offer the opportunity to improve the relationship between the curriculum in action and national curricula specifications. It offers greater confidence that those who pass science assessments have actually mastered defined capabilities rather than simply performing better than some of their peers.

Third, a number of the countries covered by the research rely on a common entrance examination set by receiving institutions to select their students for post-secondary level. Keeves (1994) notes a trend away from the use of formal examinations for the regulation of transition from secondary to post-secondary education institutions in systems where the majority of students complete the secondary cycle, in favour of such entrance examinations. This might reduce the adverse effects of examination 'backwash' on teaching and learning in science. However, the underlying reality in many developing countries is that access to higher levels of science education is heavily constrained, selection will remain necessary and the separation of school leaving certification from entrance examinations is a higher cost alternative than where a dual purpose examination is retained. It is also likely that even where school certification is decentralized its character will be heavily influenced by the requirements of entry examinations for post-school courses.

3. Assessing practical work

In the majority of countries participating in the IIEP survey, practical examinations were not common. Most science assessment at lower secondary level is through multiple-choice papers and traditional written examinations, although there is a wide variety in school-based testing between countries. In a minority of cases, science experiments conducted by pupils are included in overall assessment procedures. In Papua New Guinea and Morocco, for example, the results of continuous assessment constitute up to 50 per cent of marks in the upper secondary examination, of which 10 to 15 per cent is allocated to experimental work. Two countries in the sample integrated practical work into external examinations: Botswana and Kenya. In these, practical work accounted for 20 to 30 per cent of the marks in science subjects (Caillods and Göttelmann-Duret, 1991).

Assessing practical work remains problematic because it is expensive and inconvenient to organize. The expense of practical examining arises partly from the need to assess 'under examination conditions' (usually interpreted as where all candidates attempt the same tasks at approximately the same time under invigilation) to provide a fair test. This requires that standard apparatus and consumables should be available. If specimens of any kind are involved, this may add to the problems since large numbers may have to be procured, quality assured, and distribution problems solved. Additional costs will be involved in setting up time needed to arrange school laboratories for examinations, which often disrupts normal teaching for substantial periods.

A further cost arises if any element of performance, as opposed to outcome, is to be assessed, since this can only be undertaken as it happens. Elements of technique, some aspects of observational skills, and design and experimental development skills can only properly be assessed by observations and through following the processes a candidate goes through. Even if this is done using checklists the process will be time-consuming.

There is little evidence that overall public examination results are influenced by the amount and quality of practical work undertaken. In other words, the amount of practical work undertaken is often not a good predictor of the overall scores of candidates in examinations as they are commonly practised. In the First International Science Study, practical work was assessed. The correlation with paper and pencil tests was found to be low. Another finding was that the fewer the laboratory facilities in the school, the worse students performed (IEA, 1991).

The most obvious explanation for the lack of correlation between science practical assessments and science achievement indicators is clear (Welford, 1990; Lewin, 1992). Most science examinations do not assess the knowledge and skills that practical work might be directed towards developing: the domains relevant to the assessment of practical activity spread across skill in performing routine laboratory tasks; observation; planning, designing and performing experiments; analysis of data, explanation and prediction. Only the most refined assessments of practical work tap most of these skills. Furthermore, assessment of many outcomes thought to be promoted by practical work interacts with more general cognitive competencies.

Another possible explanation is that it is not that practical activity itself is unimportant to achievement, it is just that it is commonly approached so superficially that for all practical purposes it is a 'minds off' activity, where the intellectual elements are largely ignored (see *Section 4* above). Last, but not least, the range of marks awarded in practical examinations is often restricted – thus most scores may fall between, say, 30 and 70 per cent, rather than 0 to 100 per cent. This partly explains why performance on practical work is unlikely to be responsible for large proportions of the variance between candidates in the overall score. The long debate about the extent to which laboratory activity is helpful in learning science and improving science achievement (Lunetta and Hofstein, 1990; Haddad and Za'rour, 1986) remains resistant to singular conclusions.

The planners' challenge presented by these observations can be organized in terms of five questions, the answers to which may inform future policy on assessment of practical work:

(i)   *Does practical examining contribute to the assessment of unique knowledge and skills?*

This rarely seems the case in practice. Much practical examining is too superficial to tap knowledge and skills that are not accessible to good proxy measurements through non-practical tests, which are more convenient and less costly. Where the results of practical assessments correlate highly with performance on written items, it is clear that practical examining is contributing little to the variance of overall scores. From a selection point of view its utility is therefore questionable. Where the correlations are low, and there is evidence that unique and valued knowledge and skills are assessed, then it may be that the costs of practical examining are more defensible.

(ii)   If practical work is not examined will teachers continue to allocate time to practical work?

Given the relative unimportance of practical assessments in most public examining systems, and the evidence that practical work is much rarer in practice than is specified in curriculum documents, it would seem unlikely that current arrangements for assessing practical work are determinants of its quantity and quality. The probability is that it is only if assessment of the anticipated outcomes of practical work can be clearly specified, more prominence is given to the results of such assessment of practical knowledge and skills in selection tests, and administrative and moderation problems can be overcome, that practical assessment could have a significant effect on practice.

(iii)   Is practical examining fair in the sense that all candidates have a reasonably similar preparation for examinations and perform under similar conditions?

It is unlikely that all candidates have similar levels of exposure to laboratory environments and the opportunities to acquire skills best developed in this context. Where those familiar with this environment are competing with those who are not, differences in performance will be strongly influenced by the amount of exposure to experimental activity. This may be a situation where school effects, based on resource endowment, are stronger than in the case of more theoretically based work.

(iv)   Is school-based examining of practical activity a suitable substitute for examining practical work under examination conditions?

If the more general objections to school-based examining can be overcome then it is almost certainly preferable to assess practical activity through school-based mechanisms than through conventional practical examinations.

(v)   *Are there most cost-effective ways of assessing practical
      knowledge and skills?*

The returns on investment in practical assessments are not impressive
(Welford, 1990). Under certain circumstances simulations may be as
effective as conventional practical activity (Lunetta and Hofstein, 1990).
It is probable that much of the thinking associated with approaches to
experimental problems can be simulated with test instruments that require
stimulus material and written answers. With appropriate material –
photographs, diagrams, charts and graphs – it may be possible to test
many practically based skills (e.g. observation, collection and processing
of data, interpretation of data; experimental design) at least as well as they
are currently tested in typical practical examinations.

4. The analysis of science assessment data

A final issue of central importance to assessment in science education
and to which we shall come back in *Chapter V* is the extent to which
analysis of assessment data can contribute to planning and policy
decisions through insights into how individuals and educational
institutions are performing. The scope for this is considerable. Where
national examining systems test students on technically valid and reliable
items, large amounts of performance data are generated each year. This
can be used to explore a whole range of questions, including what
differences exist between rural and urban students, the magnitude of
gender-based differences in performance, and which aspects of the
science curriculum are consistently found difficult by different groups of
students.

Part of the IIEP research programme began to explore these issues in
one of the country case studies – Malaysia (Sharifah Maimunah and
Lewin, 1993). A brief summary of some of the findings serves to indicate
the potential of this kind of work.

First, the research showed that there were large differences in the
performance of urban and rural students at both Form 3 (lower secondary)
and Form 5 (upper secondary) levels. Only 13 per cent of rural schools
exceeded the national average in the percentage of Distinctions awarded
at Form 3, whereas 63 per cent of urban schools scored above this level.
At Form 5, in more than 80 per cent of the rural schools, students only
achieved bare pass grades. No rural school scored above the national
average for Distinctions in general science, the subject taken by those
who do not specialize in science. Interestingly, schools which performed
well tended to be larger, have larger class sizes, and were more likely to

be double-session schools (194-196). This reflects the fact that such schools usually have excess student demand and are over-subscribed. Schools with the smallest class sizes in science were rural schools with the lowest levels of achievement.

Second, analysis of test results indicated different patterns of achievement between the various science subjects. On a sample of over 5,000 students, physics had the lowest mean raw score (19.7) and biology the highest (24.9). When the sample was split into urban and rural groups, the differences in score tended to be largest in physics and least in biology. In general science, which is taken by the great majority of upper secondary students, average scores in different types of schools were low, suggesting that there was little difference in the effectiveness with which the subject was taught between schools.

Third, attempts were made to identify schools which were improving and deteriorating. This proved problematic. The performance of schools from year to year was often inconsistent. More particularly, performance in one science subject did not necessarily change in the same direction as performance in other science subjects. There also seemed to be little correlation in performance in science between one level (Form 3) and the next (Form 5), perhaps because of the selection that took place and the significant transfer of students that occured after Form 3. The analysis drew attention to the need to devise indicators of performance for schools that could be used to monitor achievement. It became clear that crude pass rates, which are currently the most common indicator used, were insufficiently sensitive to provide all but the most general indication of performance (see *Chapter V*).

Finally, the analysis of performance was extended to include item analysis. This was undertaken to explore in more depth the differences that emerged between groups of students. Using a technique developed for the purpose it was possible to identify items which strongly differentiated between high and low scoring students and urban and rural students, and between boys and girls on the integrated science examination at Form 3. From this analysis it became clear that most of the differences in performance observed were arising from a small number of items chosen to test performance. If a different set of items had been chosen, then the differences in performance would have changed considerably.

High-scoring students gained most advantage from questions which used special terminology, referred to problems involving several variables, and which referred to experiences gained from laboratory work. In general, rural students performed poorly. However, rural students did surprisingly well on some items and even performed at levels comparable to urban students. These items were often ones using a context familiar to rural students. If the test included more of these items, the gap in performance between rural and urban students would diminish. *Section 3.5* also shows how differences in performance between girls and boys may be coming from some areas of the science curriculum and not others.

From the discussion in this section it is clear that detailed examination analysis offers insights to the educational planner that can help identify schools with abnormally poor performance in science, and isolate areas of the curriculum that may especially disadvantage different groups (rural students, girls, the less able), and suggest targeted interventions that may lead to improved science achievement.

The results of detailed analysis of performance have considerable potential to improve the quality of learning and teaching providing the results are placed in the public domain and are used by those with advisory and curriculum development responsibilities; at the same time, examination 'backwash' could and should be used positively to direct attention to desired learning outcomes and areas of special weakness (Dore, 1976; Keeves, 1994).

Section 6:   Language issues in the provision of secondary science education

The choice of the language in which science is provided in secondary schools has been debated in most countries where major international scientific languages are not the medium of instruction. Scientific research and technological application progresses rapidly; published accounts of developments are presented predominantly in languages of metropolitan powers, and increasingly in English out of preference (see Eisemon, 1992). The use of other languages for science instruction, especially those without an internationally accessible scientific literature, is argued to create barriers to communication with the international scientific community. However, some research on learning and achievement – in science as well as in other subject matters – seems to indicate that children tend to learn more easily at school and perform better if the language of instruction and learning used is their mother tongue. Different options as regards the language of science provision at secondary level have been taken in the various countries surveyed. They are to a large extent related to the countries' respective history and constraints, and of course depend on other considerations than their impact on science education. Some of these implications are explored.

136

1. Provision of secondary science education in a second language

Secondary science (and other subjects) are taught in a non-indigenous foreign language in quite a few developing countries in Africa and Asia. This is the case in all the African countries covered by the research as well as in Papua New Guinea. In these countries, English or French are used as the medium of instruction at post-secondary level; in some of them, the foreign language is the medium of instruction from the very beginning of primary education. In most, it is the language of secondary school.

The use of a major international language for secondary science education has certain advantages:

> *First*, secondary school leavers need to have proficiency in the language required for following science studies at undergraduate or postgraduate levels in local institutions or abroad where access to an international language is almost certainly necessary. In practice this is not always the case. It might be the reason for student drop outs and poor science achievement.
>
> *Second*, science education in a major foreign language at secondary level facilitates access of secondary students to information and literature on recent developments that is almost impossible to provide in local languages in volume and variety. Although textbooks may be available in national languages, library and other reference books often are not.
>
> *Third*, standard terminology that is widely recognized can be employed. The use of languages with a restricted science vocabulary requires the invention of new terms often unfamiliar to most native speakers.

However, language policy which uses a foreign language has its disadvantages. Where access to secondary education in general, and to prestigious schools or science streams in particular, depends on children's proficiency in a second language, groups of students might be selected because of their language skills and not because of their scientific capabilities. Results from research on learning and testing in a second language suggest that often it is the language that is predominantly being tested rather than the skills in science. In Botswana, pupils' achievement in science, mathematics and other subjects is closely related to their proficiency in English (Fuller et al., 1994). Results of research on science learning in Kenya and Burundi (Eisemon, 1992) also indicate that when students are tested in their mother tongue they generally show higher performance levels; in the case of Burundi it was noted that the performance of the most able students was particularly poor when estimated by testing in the second language (French). It may also be that students using a second language from primary level who have not acquired basic concepts in their mother tongue may accumulate learning difficulties in science (and other subjects) at subsequent levels if they do not have fluency in the medium of instruction. As Lewin notes (Lewin, 1992), mathematics and science instruction in a language which is not the mother tongue tends to engender misunderstandings and communication problems and eventually impedes integration of science knowledge and cognitive skills because both pupils and teachers do not completely master the language system used.

It is an open question as to what extent some language problems could be ameliorated by better teacher training and design of appropriate textbooks and other materials based on understanding of language issues. In most of the countries that use foreign languages for instruction, teacher training does not formally recognize that most students have an incomplete grasp of these languages. The development of teachers' skills pays no special attention to this critical area and advice on how to cope with classes of very mixed language ability is lacking. Text material is often also only available in the foreign language, despite the reality that this denies meaningful access to many students. Rarely are bilingual materials created which offer simultaneous translations of key sections.

2. Science provision in the mother tongue across all levels of formal education

In Korea, Malaysia, Thailand, Japan and for a majority of students in the Latin-American countries studied, secondary science is provided in the mother tongue (or at least in a national language). In the South-East Asian countries mentioned above, vigorous policies have been adopted to localize the language of scientific training at all levels (major expectations are Singapore and Brunei). This has probably contributed to raising the performance of students in general since it makes science studies more accessible to all. It may be that the positive effects of science training in the national language(s) for the majority outweighs the linguistic handicap that may be imposed on scientific élites.

Developing countries that have placed particular emphasis on the development of national languages for educational and scientific training have experienced difficulties in using these languages at higher levels, especially in engineering and the physical sciences. In Malaysia, special problems were noted in relation to language competency of graduates in science which hampered access to post-graduate study and jobs at the professional level, which increasingly require fluency in English. The country is even considering reintroducing English as a medium of instruction for scientific studies at higher education level. In all cases, special support for the transition from mother tongue to an international language would seem to be a valuable complement to the use of the national language at secondary level, where access to an international language is required.

Interestingly, assessment of foreign students in American universities (at the beginning of the 1980s) showed that a large proportion of Asian students – who had not studied in the English language before – scored at a very low level in the university entrance test in English (TOEFL), when compared to students from English-speaking Africa (Eisemon, 1992). The first group apparently overcame this disadvantage quite rapidly in science, mathematics, engineering and economics since their grades at the end of the first year and their scores in the Graduate Record Examination tests of general quantitative and analytical abilities are well above those recorded for their counterparts from Africa and the Middle East. The policy implications of this are not self-evident, however; it casts doubts on the effectiveness of science learning through poorly understood foreign languages in secondary schools.

Introducing science concepts in local languages, especially at primary level, has many obvious advantages. Conversely, access to an international language is almost essential for science-based professionals training above school level. The key questions are therefore:

(i)   when to introduce a transition towards using a foreign language; and
(ii) how to manage the transition most effectively.

3.   'Mixed' instructional language policies: the case of Morocco

Some countries have opted for different languages of instruction at different levels of the education system.   Morocco's experience constitutes an interesting example of the effects and implications of a language policy which combines science education in the national language (Arabic) until   the   end of secondary education, with the provision of post-secondary science programmes in French (the predominant foreign language in this country).

In Morocco, primary and secondary education were gradually 'arabicized' during the 1980s. The first cohort of secondary school leavers which received its entire school education in Arabic entered post-secondary education in 1990. The IIEP survey revealed that substantial numbers of science teachers at lower and upper secondary level (34 and 41 per cent respectively) have difficulties in managing the transition from French (the language in which they had received their own training) to Arabic as the main medium of instruction. A majority among them still use French technical terms in their teaching and even use French 'from time to time' as the means of communication in their courses. Despite these difficulties of 'adaptation', there appears to be wide agreement that the use of Arabic has improved the understanding of science among pupils; more than one out of five (22 per cent) science teachers consider that the change of medium of instruction has enhanced pupils' comprehension of subject matter.

Students have difficulties now in managing the transition from Arabic to French when they take up science studies at post-secondary level. Many of them seem to be opting out of science studies because of the language problem. Repetition rates at the end of the first year are very high for those enrolling in post-secondary science courses. Admittedly, these were high before the change in the language policy. The results of the analysis of *baccalauréat* scores (1986 and 1990) are worrying (Caillods, Göttelmann-Duret et al., 1997): many secondary students obtaining their *baccalauréat* with specialization in 'natural sciences' have very low scores in French; some 50 per cent of them scored below average in 1990 (around 60 per cent in 1986). The percentage of those graduating from the 'mathematics/physics' stream who scored below average in French was considerable as well.

At the beginning of the 1990s, the Moroccan Government decided that all upper secondary schools should provide all students specializing in science a total of three hours of mathematics and science per week in French with a view to facilitating their transition to post-secondary programmes. The total number of hours devoted to mathematics and science activities in French were later increased to six per week. Two hours of 'translation' of science texts have since been added. Whether this is an appropriate remedy to rectify language weaknesses remains to be seen. A majority of the science teachers interviewed (62 per cent) indicated that they mostly devoted these extra periods to teaching scientific terminology in French, and that students generally showed low levels of interest and achievement in these sessions. Another alternative considered was to improve the teaching of French as a foreign language.

A few concluding remarks

- Language policy for secondary science education is determined within the national language policy on education, which curtails the options realistically available.
- Language-related learning problems in science are likely to be greater where second languages are used as the medium of instruction. Where possible it is probably preferable to introduce basic science in the mother tongue if this is widely spoken nationally. This should provide access to at least some science to all children. It may imply substantial investment in appropriate learning materials. Where the medium of instruction is a local language from grade 1, some consideration should be given to bilingual science learning materials.
- At some point during the secondary cycle, access to an international scientific language should be encouraged. In some countries this will mean retaining the local language as a medium of instruction and teaching English or another foreign language with a view to making use of it in science education; in others, the medium of instruction may change to a foreign language. Special arrangements may be needed if there is a transition.
- At higher education level, access to an international language is virtually essential for science professionals. Methods of ensuring this have to build on language competency established at secondary level.

Section 7: Training of science teachers

Comprehensive recent reviews of research on science education in the developing world come to the conclusion that what students actually learn in science depends significantly on the availability, adequate deployment and efficient use of skilful secondary science teachers. In many countries, this major condition of successful science teaching and learning is not met or is satisfied only to a very limited extent.

(a)   Who enters science teacher-training courses?

In many of the countries surveyed qualified secondary science teachers are now available in sufficient number. However, most African countries continue to be characterized by problems of an inadequate supply of qualified applicants, and output of trained teachers insufficient to meet demand from the secondary school system.

Attracting a sufficient number of science graduates into the teaching profession remains a serious problem where alternative career opportunities exist, and where graduates are far better paid in the private sector (at the beginning of the 1990s, for example, science teachers in Thailand earned six to seven times less than their graduate colleagues in the private sector). In many countries, entering the teaching profession appears to be the result of a negative choice by many students – i.e. the last option available. This seems to be less of a problem where the output of those with science qualifications has been great (for example, Korea, Morocco) and there are more applicants than places for teacher training. A surplus of science qualification holders, whether as a result of an over-production of graduates (Morocco, Jordan), or of a shrinking labour market (Senegal), means that faculties and colleges of education do not have difficulty in recruiting and selecting candidates.

The profile of science teachers in different countries varies according to the educational levels/credentials that are required for *admission* to specialized teacher-training institutions or university faculties in which future science teachers are trained. These are generally equivalent to the entry requirements for a science faculty. Some countries are, however, slightly more selective. Such is the case of Morocco, where admission to teacher's college and to the regional pedagogical centres is also conditional on success in an entrance competition. In view of the shortage of science teachers, other countries are considering lifting certain conditions. In Botswana, for example, the high wastage rate in the Pre-Entry Science Course (PESC) is said to contribute to the persistent shortage of local science teachers in the country, hence there is a debate on whether it should be maintained. In Kenya, where there are not enough secondary graduates with the required level in science, it has been necessary to lower the minimum score for admission in order to attract students into science education (other than medicine and engineering). Despite this, the number of teachers trained in science continues to decline in relative terms. In response to an exceptional problem of teacher supply in science, South Africa has developed a scheme whereby students from disadvantaged backgrounds who under-performed at matriculation at the end of secondary school, and thus who would not normally qualify to continue, are selected and trained in a sandwich type of course with a large on-the-job component. In some countries it is not necessary to have succeeded in any examination to enter the science teacher-training colleges (this is the case in Thailand, the essentially private professional institutes of Chile, or the post-secondary teacher's colleges of Argentina).

Thus, future teachers trained by these institutions are chosen less selectively.

Effective science teaching requires not only a sufficient level of motivation and general education, but also sound subject-matter knowledge, the pedagogical and didactical skills required to transmit science knowledge effectively to children, and practical experience with pupils' learning difficulties and possible ways of addressing them (see Avalos, 1995; Keeves, 1991; Lewin, 1992; Haddad and Za'rour, 1986; Walsh and Gregorio, 1994; Ware, 1992a and 1992b; Walberg, 1991).

In many parts of the world, there is serious concern about the shortcomings of existing pre-service training programmes in 'producing' science teachers who can meet these requirements. It is an open question as to what extent in-service training can compensate for the deficiencies of initial training. This leads us to consider a discussion of developments regarding the length, organization, content and modes of pre- and in-service training of science teachers.

(b)   Organization, content and modalities of teacher training

Training of science teachers presents many different patterns in the countries investigated. Some countries still distinguish between two categories of teachers – those destined to become lower secondary teachers, and those who are qualified to teach in upper secondary.

- *Lower secondary teachers* are often trained in two or three years after secondary school in regional pedagogical centres or teacher's college (Morocco, Thailand, Papua New Guinea, Senegal). Alternative routes exist and students can be offered one year of pedagogical training after two years at the university (Morocco, Senegal and Burkina Faso).
- *Upper secondary teachers* are trained in four to five years after secondary school in teacher's colleges (e.g. Morocco recently, Thailand) and teacher-training institutes (Argentina, Chile). An increasing number of countries have moved to training secondary science teachers through programmes combining pedagogical training with a science degree course (three to four years of study at a faculty of science), or through a one-year postgraduate teacher-training course after obtaining a science degree.

Whatever the route into teaching, secondary initial teacher-training programmes are becoming increasingly three or four years in length, with

143

no difference being made between lower and upper secondary science teachers. However, as Ware points out (Ware, 1992b), the length of teacher-training programmes seems to be a matter of opportunity or custom rather than of proven value. Long programmes are costly for both teacher trainees and those who provide them (for the most part, public authorities). There is also no evidence that longer programmes have better results. Much obviously depends on the content and organization of science teacher-training programmes.

A major curriculum issue to be addressed relates to the respective role and length of *theoretical pre-service teacher education,* on the one hand, and *practical on-the-job training,* on the other hand. Effective science teaching undoubtedly requires an adequate level of subject-matter knowledge and subject matter-related didactic skills which are unlikely to be acquired efficiently on the job. A review of relevant research on recent experiences with science teacher training carried out by Avalos (1995) concludes that teaching practice is the only effective way for teachers to actually grasp pupils' approaches to science learning, learn how to manage a class, organize work and student monitoring efficiently and learn other practical skills, without which a theoretical knowledge base cannot become fully effective. Furthermore, novice teachers are reported to be more quickly effective to the extent that they receive adequate support – through experienced and well-trained master teachers (mentors), teacher guides and support materials, etc. Unfortunately, the 'practice' component of most secondary teacher-training courses – especially those provided by universities – tends to be very short (varying between 12 and 24 weeks according to Rosier and Keeves (1991). Furthermore, the research found some evidence that support services for novice science teachers still constitute a weak point in many systems. Even in Morocco, where the inspection of secondary teachers is staffed and functioning at a satisfactory level, the survey among secondary science teachers revealed that inspection and advisory visits are rare over the first years of service.

Science teacher-training programmes are commonly composed of 'science content', pedagogical instruction including supervised teaching practice, and general education courses. The balance of these three components varies considerably from one country to another and sometimes from institution to institution within a country, as *Table 3.2* shows. The science component is much stronger where science teachers spend most of their training at a faculty of science; the respective part of the 'education' and 'general education' component tends to be higher in college-based teacher-training programmes.

Table 3.2   Time spent on science versus general education courses on teacher/learning programme

Country		Science %	Education %	General ed. %
China		70		
Egypt		80	20	
Hungary	(LS)	50-55	15-20	30
Indonesia	(D3)	64	25	11
	(S1)	66	23	11
Nigeria	(US)	47	40	13
Philippines	(LS)	12-25	19-21	60-64
	(US)	30	22	48
Thailand	(US)	52	27	21
Venezuela	(LS)	35		
	(US)	60		
USA		40	21	39
UNESCO Model		50	35	15

Source: Ware, S.A. 1992b.

Avalos (1995) comes to the conclusion that faculties of science are in most countries the place where secondary teachers can acquire the highest possible level of subject knowledge. However, the programmes they offer often do not include important topics – such as integrated science, environmental issues or technology – which science educators may wish to see become an integral part of science teacher training. Moreover, since they are not geared towards the special needs of prospective science teachers, such courses do not generally teach how to teach science, i.e. how to make scientific knowledge accessible to a mass school population.

Pedagogical universities or teacher-training colleges seem to be in a better position to tailor their programmes to the specific needs of science teacher training and also to respond to possible curriculum changes. These institutions, however, tend to offer programmes in which the pedagogical procedures are predominant, often at the expense of the provision of quality subject-matter knowledge. The recommendation of many science education specialists is to maintain the science knowledge component of teacher training within the Faculties of Science and to provide only some complementary pedagogical training in Faculties of Education or teacher-training colleges. This option also tends to be less costly than offering the complete cycle of science teacher training in teacher-training institutions, since the latter are often residential and – due to lower enrolment levels – not in a position to operate the same economies of scale as universities. However, the structures and curricula of science faculties may prove to be too rigid to adapt themselves to 'concurrent' (parallel) provision of theoretical and classroom-based practical training. In practice, therefore, it is more common to find sequential science teacher-training programmes composed of a science degree course, followed by a course concentrating on theoretical and practical aspects of subject pedagogy. However, the IIEP's in-depth case study on Morocco and Avalos' review (1995) of literature on science teacher training concludes that the widespread 'Pedagogical Year' (added to a Bachelor of Science degree), in its predominantly theoretical and institution-based form, may be insufficient to generate the teaching skills and attitudes required for effective science teaching. Appropriate support after initial qualification could help overcome any such deficiences.

According to the constructivist view it is the use of certain instructional approaches and methods that can help to greatly enhance the effectiveness of science teaching and these need to be included in preparation for science teaching. Avalos (1995) as well as the authors of a state-of-the-art analysis of science education in French-speaking countries (Giordan and Girault, 1994) mention in particular 'reflective approaches' (reflection on the students' and student teachers' own concepts and views of science and scientific phenomena, reflective exploration of teaching strategies and difficulties experienced during practice teaching, etc.) and micro-teaching (simulation, monitoring and critical analysis of teaching organized in small groups of peer students). In some of the countries included in the IIEP project, science teacher-training programmes have recently been developed along these lines. A particularly interesting example (reported by Avalos, 1995) is presented in *Box 3.1*.

Box 3.1   Science teacher training at the Osorno Professional Institute, Chile

The Professional Institute of Osorno (one of the smaller universities in the South of Chile) is in the process of developing a new curricular structure for secondary teacher training. It attempts to produce integration between content and pedagogy from the start of the course of studies. Furthermore, the traditional course pattern has been radically altered, moving away from discrete courses in the area of Methodology and Foundations (psychology, sociology, philosophy) with field experience and practice teaching taking place towards the end of the programme towards a series of thematic seminars that run concurrently with the subject specialization. These seminars involve not only 'learning about' and critical reflection but also a practical component and evaluation of student teacher learning through written reports and oral examinations. This curriculum structure has been fully implemented for mathematics teacher training, while its application to the training of science teachers is under preparation.

Source: Avalos (1995).

Another crucial question that needs to be addressed is how science teachers are prepared and how they could be prepared for undertaking practical work. Little research has been done on this point. The analysis of science teacher training in Morocco carried out by the IIEP suggests that very little, if any, 'hands-on' individual laboratory work is done during the first two years of degree courses at the Faculty of Science – which tends to become the 'common core' of science teacher training in this country. The time given to training for practical work also seems to be restricted in the additional professional training they receive.

It is often suggested that these and other shortcomings of pre-service training could be addressed through in-service training programmes (INSET) for science teachers. The IIEP international survey points to the irregularity and mostly very short duration of INSET for science teachers. Avalos (1995); Walsh and Gregorio (1994) come to the conclusion that non-conventional forms of in-service training (e.g. distance teacher training or 'sandwich' courses combining institution-based and on-the-job training) are possible means of enhancing the level of subject knowledge of science teachers. Teaching practices on the other hand may be more difficult to change through in-service training. According to Avalos (1995), programmes of collaborative research between universities and secondary science teachers (bringing together university academics and practicing teachers for the sharing of ideas and the development, experimentation and evaluation of new teaching-learning approaches)

147

may be more effective and attractive in this respect, wherever it is possible to organize these. It may be even more important to provide support for novice science teachers, especially through structured opportunities for interaction with other novice teachers, experienced teachers, university-based instructors/teacher trainers, etc., and through appropriate teacher guides and materials.

(c)   Improving the cost-effectiveness of science teacher training

Three to four years of pre-service training, the predominant way of training science teachers may be one of the most expensive ways of providing initial training to teachers. Rapid enrolment expansion at secondary level has thus led many countries to look for less expensive ways and means of addressing the teacher shortage. Crash courses, distance teaching and in-service training in general can constitute an alternative to costly forms of pre-service training. Little is known, however, about the cost and effectiveness of different methods of teacher training.

A comparative analysis of the cost-effectiveness of different forms of training experienced by underqualified primary mathematics teachers was carried out in Sri Lanka. It compared training through: (i) distance education; (ii) pre-service education at the colleges of education; and (iii) in-service education at teacher's colleges and came to the conclusion that the distance-learning programme represents one-sixth of the cost of the pre-service training programme and one-third of the conventional in-service training; yet, pre-service training at the colleges of education turned out to be the most effective in increasing the achievement levels of teachers in mathematics knowledge and skills; distance in-service training proved to be less effective than conventional pre-service training, but slightly more effective than the (more expensive) campus-based in-service training (see Nielsen et al., 1991). More studies of this type are required. The question of the cost-effectiveness of different in-service and on-service programmes is taken up again in *Chapter IV*.

A few concluding remarks

- Efficient ways of improving the subject-matter knowledge of science teachers need to be found, especially in those countries where difficulties in recruiting science teacher trainees have led to the lowering of admission criteria. Training at Science Faculties may be a more effective and less costly way of imparting decent levels of subject-matter knowledge than courses provided in special Teacher-Training Colleges.

- Teaching practice is an essential training component through which indispensable didactic and classroom-management skills can be most effectively acquired. Increasing the share of on-the-job training can be considered as a way of enhancing the effectiveness and reducing the cost of science teacher training.
- Novice science teachers need advice and support. Local or school-based systems of support (including master teachers, heads of departments, inspectors) need to be further developed in order to address this challenge.

In this chapter, a number of issues crucial to science education provision have been discussed. The best solution for any one country very much depends on its own particular context, history and financial constraints. A few recommendations nevertheless have been made at the end of each section on specialization, curriculum and assessment issues as well as on the training of teachers.

Planners are also concerned with the most effective ways of dealing with these issues. This will be discussed in *Chapter IV*.

Chapter IV

Cost-effective approaches to science education: an analysis of issues

This chapter falls into four main sections. The first looks in detail at the attractions and disadvantages of sustaining special science schools within public school systems. The second explores the cost implications of laboratory provision and science kits. The third reviews some evidence on the development and delivery of science learning materials. Finally, the range of options in supporting and managing school science departments more efficiently is considered.

From the data which were presented in *Chapter II* it is clear that the unit costs of providing science education at secondary school level are amongst the highest of all school subjects. This makes it all the more important that the science education that is provided is efficiently managed, results in real learning gains, and is designed to nurture scientific skills and capabilities which contribute to national development in a way which balances the needs of the majority with the needs of those who will follow science-based careers.

Planners of science education have a special interest in the selection and tracking of students into science since the costs of providing places are usually higher than for other subjects. If access to high-quality facilities cannot be afforded for all, specialization and selection have to occur. One response to the need to increase flows of well qualified science-based school leavers at affordable levels of cost has been the introduction of special science schools in a number of countries. In these, science continues to be taught in normal schools with more modest resources. The next section explores some evidence on the effectiveness of this strategy in Malaysia and Nigeria.

High laboratory building costs are often the result of standards of design and equipment that mirror those in industrialized countries. The question must be raised as to whether this is justified by pedagogical necessity, common patterns of use, demonstrated learning gains, and the ability to support running costs. There are obvious attractions to utilizing lower cost methods of delivering practically based learning experiences through deployment of science kits and the greater use of demonstration teaching methods. Where the supply of good quality science learning material remains problematic (often despite several decades of curriculum development), the reasons for this need careful examination. Issues concerned with the design of practical facilities and the provision of learning materials are discussed in the second section below.

Organizational changes and effective support mechanisms for teachers have been undervalued in much of the literature on science education reform. Recently, more attention has been paid as to how existing structures can be modified to improve the quality and quantity of science education available, monitor and support the efforts of schools and individual teachers, and recognize that science departments may require special management and organizational skills which are different to those needed for other subject areas. This is the subject of the final discussion in this chapter.

Section 1:   The case for and against special science schools

The provision of secondary science education involves a number of strategic decisions concerning the most effective modes of delivery. Central to the choices open are decisions on whether to provide largely undifferentiated science curricula to all students or whether to introduce various levels of specialization. A number of countries, such as Nigeria (Kano State) and Malaysia have introduced special tracks for science students in the form of special science schools.

Several different arguments have been put forward for the case that can be developed in favour of teaching science in special institutions at secondary level.

First, as noted above, good science teaching is expensive and the necessary resources are scarce (competent teachers, well-equipped laboratories, up-to-date libraries). It is therefore justified to concentrate on providing quality science education to a small minority of the best students in special institutions where a well resourced and focused science education can be provided.

Second, an adequate supply of competent high-level scientists and technologists is needed to fill places in further and higher education in appropriate disciplines, and to contribute subsequently to the development process. This may be better assured through tracking the best science students into special institutions than through a dilution of effort across the whole school system.

Third, it is sometimes argued that scarce scientific talents may be best developed on the basis of early and extensive opportunities to learn about science. Scientific thinking may be more difficult to impart to older students.

Fourth, special science-focused institutions with special admissions policies may be needed to increase the participation of historically marginalized groups in science- and technology-based education and employment.

The first proposition has most force in the poorest countries, where there is no possibility of most secondary-age children being enrolled in schools with adequate facilities and trained science teachers. If national needs for science-qualified school leavers are to be met, some concentration of resources seems inevitable. Without it the dilution of effort is likely to be such that few students qualify at acceptable standards. Middle-income countries can probably afford to resource the teaching of science in all secondary schools providing realistic decisions are made about the range and type of equipment required and the rate of replenishment that is reasonable. In these cases, the issue is about what gains there may be from special provision in comparison to the benefits from enhancing normal provision.

The second argument applies particularly where there are shortages of science and technology-trained nationals in the labour market and where post-school opportunities in science and technology are expanding at a greater rate than the output of science-qualified school leavers. Both of these circumstances suggest the need to raise the output of science and technology-qualified school leavers at secondary level. One approach to this will be to expand specialized provision and increase incentives to enrol in special institutions.

The third proposition presupposes that there is something unique about the intellectual activity of science that requires early systematic exposure if creative and competent scientists and technologists are to be trained. The evidence for this view is mixed, as discussed in *Chapter III*. Specialization does seem to lead to higher levels of achievement in science amongst students (see the results of the Second IEA Science Study on upper secondary students), but this is hardly surprising. This evidence alone does not lead to the conclusion that early specialization is essential. Whether and how high levels of achievement might carry through in some meaningful way into performance in employment remains an unanswered question. Rates of cognitive development amongst children appear to place some limits around how early conceptual structures and abstract analytical thinking can be developed. On the other hand, there would seem little merit in delaying the intellectual challenges of science until late in the secondary cycle.

Lastly, there are cases where special institutions have been established to teach science to redress historic imbalances in participation and achievement. This may be an effective strategy if it is coupled with a consistent maintenance of standards within the context of a positive discrimination policy for a marginalized group.

The arguments advanced in favour of special institutions to teach secondary science have to be balanced against difficulties that often arise in establishing a separate track through secondary education for science students. These are likely to include:

- establishing the reliability and validity of methods used to select students suitable for special science schools;
- the possibility of undermining science teaching in the normal school system;
- the difficulty of assessing what level of additional costs is justified;
- the risks of academicizing secondary science in élite institutions to the extent that it loses relevance to national development needs;
- the possibility of overshoot from deficit to surplus in supply of science-trained school leavers once initial demand has been satisfied;
- the probability that without separate examination arrangements, approaches to learning and teaching science will simply mirror those employed in ordinary schools and lack enriching experiences.

Each of these potential difficulties is worth some discussion. The first implies the need for technically sophisticated selection methods which can be shown to be valid and reliable. A decision also has to be taken as to when such selection should occur, bearing in mind that the earlier this happens the greater the likely range of error associated with a given result (if only because of the increased probability of cases of 'late development'). The fewer selected and the greater the costs of their education, the more important it will be to have confidence in the selection process. The probability that some who are selected may subsequently prove not to have the attributes to benefit from special schooling must also be recognized.

Second, there is a real possibility that, where a significant number of special science schools are established, they will undermine science teaching in the bulk of the schools. This may occur for several reasons. The most energetic, creative and qualified teachers will be attracted to science schools; spending on science facilities and equipment may be diverted into special institutions; the best science students will be removed from normal schools where they act as role models for others; parents and the community may wrongly assume that science cannot be taught properly except in special institutions with a special concentration of resources.

The *third* problem recognizes that it is one thing to accept the principle that special science schools need to be more expensive than normal schools; it is another to decide what an appropriate ratio might be. Staffing ratios may need to be more generous – if the normal pupil teacher ratio is 25:1, should it be allowed to fall to 15:1? 10:1? Or even less? Similarly, it is possible to envisage escalation of equipment costs to very high levels if every request with a possible justification is to be met: state-of-the-art analytical instrumentation, personal computers for data analysis, individual work space, etc. There will be a point of diminishing returns where higher costs have little effect on science achievements. Expenditure may need to be controlled by those without a stake in an expanding budget if appropriate levels are to be set.

Fourth, no growing economy employs more pure than applied scientists – teachers aside. Technology is what transforms the environment and increases productivity and this is based in large part on the systematic application of knowledge and skills with a scientific basis. Thus, the need for science-qualified school leavers is likely to be skewed in favour of those with a sound basic science education and the rudiments of applied science skills. There is a risk that the prestige associated with pure science in many education systems, and the preferences of teachers (a minority of whom have usually been trained as applied scientists), may reinforce each other to emphasize the theoretical at the expense of the practical, the abstraction at the expense of the application. It may therefore be important to stress technological and applied science in special institutions, rather than pure science, if science graduates are to play a full and cost-effective role in the development process.

Fifth, the proportion of those enrolled in a special science track is an important consideration. This is unlikely to be large if special institutional arrangements are to be made. Cost and resource implications will determine that relatively small proportions are enrolled. Over time the supply and demand for science-trained school leavers will change. The larger the commitment to a special track, the greater the likelihood of overshoot from shortage to surplus once the demand for a basic stock of science-qualified workers has been satisfied.

Sixth, if special science institutions follow national curricula, and particularly if they prepare students for the same national examinations, it is possible that they will stress the same pedagogic styles as are found in successful ordinary schools. It would be surprising if they did not. It may be that with the best students, creative teachers, and extensive resources, time will be found for the more exploratory, challenging and intellectually demanding activities that can stretch understanding. However, schools are likely to be judged publicly by their examination record and much effort may be directed towards maximizing results. If special schools are to emphasize learning goals that are different and more extensive than those associated with the teaching of science in ordinary schools, assessment practices may need to be refined, or additional more challenging assessment provided.

1.   An analysis of two examples of special science schools

In Malaysia, three types of schools are particularly well resourced to teach science. The MARA Junior Science Colleges (MJSC) and the Science Residential Schools (SRS) were established in the 1970s; in addition, the Fully Residential Schools (FRS) include a number of well-known selective schools established under the former colonial power. The *Mara Junior Science* Colleges were set up with the main objective of helping to increase the participation of 'Bumiputra' (i.e. ethnic Malays and other indigenous people) in science- and technology-related fields. MJSCs are not controlled by the Ministry of Education but by the Secondary Education Division of MARA, a development agency under the Ministry of National and Rural Development. Entry into the science stream requires good passes in mathematics and integrated science in Form 3 and priority is given to candidates from rural areas.

From 1986 to 1990 enrolment increased from 7,860 to 8,400, representing a little over 2 per cent of total enrolment at Form 4 level. Males outnumber females in the intake. Since 1988, the intake has been entirely at Form 4 level (although originally Form 1 students were admitted). The MJSCs offer courses not found in other schools (e.g. computer literacy and thinking skills), appoint their own teachers, and follow their own teaching programmes. However, students sit for standard national examinations.

The *Science Residential Schools* (SRS) are state schools with a special role in increasing participation and achievement in science education amongst rural and educationally disadvantaged students. These schools have an intake quota of 70 per cent rural and 30 per cent urban students. There are also quotas for each state (based on numbers taking the scholarship examination) and for students from different types of school. The major intake into these secondary schools is at Form 1 level, with some small additional recruitment in Form 4. In 1989, 91 per cent of upper secondary students in the SRS were in the science stream. Slightly less than 1 per cent of all Form 1 enrolment is admitted to SRS, with a majority from the more rural states. Nearly two-thirds of the intake is male.

Fully Residential Schools (FRS) differ from Science Residential Schools in that they admit students on the results of the primary scholarship examination in Standard 6 and do not apply quotas for rural students. The majority of the intake is in Form 1 and small numbers join in Form 4 if they have exceptional Form 3 results. In 1989, there were a total of 4,000 students at the upper secondary and sixth form level in FRS, representing about 2 per cent of the total enrolment at these levels. Of these, 61 per cent were in the science stream. Male participation exceeded female by 17 per cent in the science stream and by 4 per cent in the arts stream. Arts students form a majority in Form 6 in these schools.

The *Special Science Schools in Kano State* (Nigeria) were established in the late 1970s to increase the flow of Kano indigenous people into the labour market who had science-based qualifications[1]. The schools were established with boarding facilities, unlike normal secondary schools. Students were selected from across the state on the basis of examination performance in Form 2. The schools were provided with eight laboratories each and appropriate equipment. The best teachers were attracted to these schools by the offer of superior conditions of service. Two girls' science schools were established alongside the three boys' schools because there was a concern to increase the number of girls studying science. By 1990, these schools enrolled a total of over 2,800 students, accounting for between a third and a half of the students reaching school certificate level and taking science subjects in the state (Adamu, 1992).

These two cases can be contrasted in a number of respects. The Malaysian school system is well resourced and science laboratories exist in most secondary schools. From a physical resources perspective there may no longer be compelling reasons to concentrate on building up equipment and facilities in a very small number of secondary schools. Good science can and is taught in many ordinary secondary schools. However, human resources may be in shorter supply.

1. Bauchi, Jigawa, Kaduna, and Katsina States have also established special science schools.

There appears to be an overall shortage of science teachers (if the goal is that all secondary science should be taught by graduates). From case study data on classroom practices in science it is also clear that much teaching falls short of providing students with challenging and exciting lessons in science. Too much teaching is ritualized and too closely bound to the syllabus to offer opportunities for exploratory learning and the development of many scientific thinking skills (see *Chapter II*). The various special science schools can offer a different climate for science learning as a result of their ability to attract and retain some of the best science teachers and students. This potential advantage may deteriorate over time if science schools lose their special character and begin to resemble normal well-resourced schools, as appears to have been happening with the Science Residential Schools.

In the Kano State education system, resources have never been as widely available as in Malaysia. At the time the special science secondary schools were established, adequate physical resources to teach science existed in very few secondary schools. This has remained a problem. Whilst it is the case that most schools cannot be well equipped to teach science to the end of the secondary cycle, it is rational to concentrate resources in the science schools rather than spread them across the whole system. Without concentration few schools if any would be able to establish and maintain adequate physical facilities. In Kano, the science schools have succeeded in attracting some of the best science teachers away from other schools. The disadvantage is that this reduces the chances of science being taught more effectively in ordinary schools. It can be claimed however that if facilities in outlying schools are very modest or non-existent, even the best-trained science teachers would be limited in what they could achieve.

2. Impact on the number of students enrolling in science

There is evidence in both Malaysia and in Kano State that the number of science-based school graduates from groups that were previously under-represented has increased substantially. Many Bumiputra students enrolled in MARA and SRS schools might not have continued to study science had they been enrolled in normal secondary schools. These schools almost certainly provide a disproportionate number of Bumiputra entrants to science- and technology-based further and higher education. However, there is evidence from the school case studies (Sharifah Maimunah and Lewin, 1993) that increasing numbers of students in these schools are beginning to favour non-science subjects when choosing options for study at levels above Form 5. If fewer

Bumiputra students continue with science-based courses, this would defeat the purpose of such schools and may reverse the trend to achieve a better balance in the labour market between Bumiputras and others.

In Kano, the output of science-school graduates seems to have been sufficient to provide Nigerians to fill many of the science-based posts previously occupied by those from elsewhere. Though unknown numbers of qualified school leavers have emigrated to other parts of Nigeria or abroad, many have entered the local labour market. Without a proper tracer study it is impossible to know the flows with certainty. However, the signs are that quantitatively much has been achieved (Adamu, 1992).

3.   Cost-effectiveness of special science schools

The costs of special science schools are not clearly separated in the available data. Some estimates can be made. Since teachers' salaries will be the largest element in recurrent costs, pupil/teacher ratios will give an indication of relative costs. Average pupil/teacher ratios (ptr) in Malaysia at secondary level are about 20:1, pupil/teacher ratio in the FRS and SRS appear to be between 10:1 and 15:1. In the MJSC school in the IIEP case-study group, the pupil/teacher ratio was about 9:1. The Mara School will therefore be at least twice as expensive as an average state school on recurrent salary costs. As average salaries in the Mara School are likely to be higher and the ratio of support staff is also greater, this represents a minimum additional cost. In fact, the special schools were much more expensive than this in the mid-1980s. The unit cost figures for 1984 were M\$694[2] for secondary schools in Peninsula Malaysia, M\$4,093 for Fully Residential Schools, and M\$7,401 for MJSC (Government of Malaysia, 1985)[3].

The critical question is whether the margin of superiority of the Residential (FRS/SRS) schools over the national average in the proportion of distinctions and credits awarded, justifies the additional costs. The best secondary schools achieve comparable examination performance to the FRS/SRS at lower unit costs. No data were available on the quality of the intake of the different schools. A fair comparison, and an answer to the question posed, can only be obtained from analysis that takes this into account. This would need a study controlling for the intake scores of

---

2.   Presumably including schools without Form 6 classes.

3.   The costs of boarding are responsible for some of these additional costs but unit costs in normal residential schools were only estimated as M\$1,172 at this time (Government of Malaysia, 1985:59).

students. Students in the MJSC schools clearly perform extremely well by any standard.

Our estimates therefore suggest that special science school provision is expensive, ranging from a minimum of about three times normal costs to more than ten times as much. Whether the additional costs are worthwhile will depend partly on the levels of performance that are achieved. The performance of students in science in the FRS and SRS schools in Malaysia was above the national average, as might be expected from the selective nature of their intake. The pass rate in the sciences generally exceeds 95 per cent, whilst the national average pass rate was between 80 and 90 per cent, depending on the subject. More indicative of the nature of differences in performance was the fact that typically the proportion of distinctions and credits obtained was greater in the FRS/SRS schools (in 1990, about 80 per cent, compared to 50-60 per cent in ordinary schools) and greatest in the MJSC (90 per cent) schools.

The Science schools in Kano produced nearly 4,400 graduates between 1980 and 1989. This greatly exceeded the total output of science-qualified leavers from normal secondary schools since the formation of Kano State in 1968. In quantitative terms, the schools seemed to have succeeded in increasing the output of qualified science school leavers. What evidence there is on performance suggests that these students' results at the secondary school examination are well above the average for normal school candidates. In the science schools, pass rates average between 55 per cent and 70 per cent compared to 5 per cent to 14 per cent for candidates from the normal school system. The students are a selected group however and there is no way of correcting for this with existing data when comparing examination performance between schools. Students in the science schools have superior scores in all subject areas. Caution is therefore needed in drawing conclusions.

Secondary pupil/teacher ratios in Nigeria average about 40:1. In the science schools they seem to be about 16:1. Non-salary recurrent costs are higher in the special schools, but detailed data are not available. It seems the first special science schools were probably costing about four or five times as much per student as the normal schools with equivalent age ranges. Boarding costs probably account for about 25 per cent of the additional expenditure.

4. Teaching practices

Case-study data from Malaysia did seem to indicate that there were significant differences in pedagogy between the MJSC school visited and other residential and ordinary state schools. The climate within this school appeared to be very achievement-orientated, with teachers under pressure to obtain excellent results from students. If they failed, there was an expectation that they would be moved to another school. It also appeared to be the case that quizzing had been institutionalized, so that there was frequent reviewing and testing of students' knowledge and abilities on virtually a daily basis, unlike in many other schools visited. In contrast, the approach used to the teaching of science in the other residential school visited appeared similar to that employed in high-scoring ordinary schools.

In the Kano Special Science School, all students have to study the three main sciences, mathematics and further mathematics. Learning and teaching activities in the Science schools are intended to reflect the best practice available. The evidence (Adamu, 1992) suggests that much teaching follows fairly traditional approaches rather than the more innovative elements of recently introduced science curricula. Practical work occurs less frequently than curriculum materials advocate. Teachers appear to place a heavy emphasis on the acquisition of the kind of knowledge needed to pass the secondary school leaving examination, and under-emphasize the development of intellectual skills above the level of recall. The examination was not changed when the science schools were introduced.

In neither country do tracer studies exist, which could show whether the graduates of special science schools scored well in higher levels of education and whether they found relevant employment. It is of consider-able interest to determine what proportions of such graduates find science-based employment locally and overseas. If only a minority continue with careers that are science based, then the relatively high cost of such institutions will be more difficult to justify than if the majority are in employment with a scientific or technological base. This may also be the case if the majority of the successful emigrate.

In conclusion, there is enough evidence to suggest that special provision of the kind found in Malaysia and Nigeria does succeed in producing students who obtain high levels of qualification in science subjects. Such institutions appear to be effective in increasing the numbers qualifying from historically disadvantaged groups. What remains difficult to judge is the cost-effectiveness of such institutions – more detailed data are needed which can be used to control for entry scores and compare achievement gains with those achieved by high-scoring ordinary secondary schools which operate at much lower cost levels. A major reason for additional costs is the amount of boarding in special science schools. The empirical question is whether the effects on achievement result from the presence of boarding or differences in learning and teaching that are unrelated to boarding. This also requires a separate study.

Section 2: Science laboratories and science kits – costs and options

As was discussed in *Chapter II*, the costs of teaching science are very high, capital costs in particular. The data from the research raise two questions for planners: why should laboratory costs be so high, and, is it possible to provide an acceptable range of equipment for practical work at lower cost?

There appear to be no data that convincingly link the costs of laboratory facilities with levels of achievement. The reasons for high ratios of cost seem to lie more in a combination of local preferences for standards of provision found in the best schools, and the availability of external finance for high-cost facilities. In neither case is this likely to result in design sustainable across the school system as a whole.

The lowest cost ratios are found where science rooms rather than science laboratories are being built, where basic services are available (electricity, water) but not for every student, and where the equipment provided is simple. These facilities should have construction costs close to those of normal classrooms and modest additional costs for suitable furniture (working tables, secure storage space). Data from Botswana suggest that such provision may be achieved at about one third of the cost of a fully equipped laboratory.

The cost of consumable materials can be a significant burden depending on how these are provided and how much practical activity is undertaken. Costs tend to be higher for chemistry if analytical quality reagents are used for classroom practicals. Biological materials are more likely to be available in the local environment at low cost. Much physics teaching does not consume material, since many things can be re-used – wire, lenses, thermometers, etc. Those things that are consumed, e.g. batteries, are often available in the local market. Various techniques can be used to reduce expenditure on consumable materials. Most obviously, measured quantities of reagents can be prepared for class use to prevent wastage, as is the case in some schools in Malaysia (Sharifah Maimunah and Lewin, 1993), and household chemicals can be substituted for analytic grade material. Alternatively, demonstrations rather than class practicals can be presented. There is therefore a challenge, particularly in relation to chemicals, to design curricula in such a way that the use of scarce or expensive consumable material is minimized.

A number of countries have experimented with the use of science kits as a response to the need to support practical activity at sustainable levels of cost. Science kits designed for use in ordinary classroom environments may offer a much cheaper route to support the teaching of empirical aspects of science than fully equipped laboratories or science rooms. They may also be of value where science rooms or laboratories exist, but are poorly equipped or inadequately maintained. The experience with science kits, has been reviewed by Ross and Lewin (1992). As from the 1960s various low-cost equipment programmes were developed alongside science curriculum development projects and these included the production of various types of science kits. More than 30 countries were deploying science kits of one kind or another in the 1980s.

The content of science kits varies widely and some examples are presented in *Box 4.1*. The examples given are of centrally provided kits which are probably the most common kind. The Institute for the Promotion of Teaching of Science and Technology in Thailand is a fairly typical example of a centrally organized facility that designs prototypes of kits and then sub-contracts their production to commercial producers (Wilailak, 1991). Teacher-made kits are also produced in some countries.

Most science kits provide a mixture of permanent, perishable and consumable items. Permanent items such as microscopes, thermometers and electric meters are the ones most likely to be imported. Permanent items can be expected to last three or four years or longer. The cost of these has probably been falling as the international market has expanded and become more competitive. Perishable items (chemicals with a limited shelf life, some biological specimens may or may not be available locally and have to be regularly replaced. Many consumable items (candles, matches, razor blades, household chemicals) can often be obtained locally but some, especially analytic chemicals, may have to be imported. The kits examined contain a surprising number of items that are widely available – e.g. electrical wire, matches, magnets (often available in old electric motors) and torches.

Box 4.1 Examples of science kits

The Hoshangabad Science Teaching Programme provides an integrated science kit for middle schools which has 44 basic items designed for use by groups of four students. These items include hand lenses, a simple microscope, a stove, compasses, thermometer, torch bulbs, slides, razor blades, naphthalene, candles, and a range of basic chemicals.

The 'mini-lab' in Sri Lanka is more ambitious and includes 82 items designed to cover virtually all the secondary science syllabus. It contains electrical meters, circuit boards, an electronics kit, atomic models, spring balances, a dissecting set, a microscope, a wide range of glassware items including burettes and lenses, thermometers, litmus paper, and electrical wire. However, no chemicals are included, it being assumed that those needed can be obtained locally. The unit is intended primarily for use for teacher demonstrations but can be adapted to support individual experimentation.

A third example is the kits in use in Papua New Guinea at lower secondary level. These kits are both more comprehensive and more expensive. The physics kit incorporates a range of electrical meters (voltmeter, ammeter, galvanometer), electricity and magnetism equipment including bar magnets, compasses, batteries, bulbs, wire, switches, connectors, an electric motor, bells, battery cell plates, solar cells, thermal expansion equipment, bi-metallic strips, mirrors and lenses optical equipment, spring balances, pulleys, standard masses, power supplies. There is an additional kit of assorted items suitable for all of the science subjects. The kits contain material intended for grades 7 to 10. Teachers have to select items from the kits for particular lessons.The kits contain sufficient equipment for class sets of about 40 students working in groups of four. Virtually all of the kit items are imported.

Source: Ross and Lewin, 1992

Many of the early kits were simply collections of apparatus with a minimum of guidance on use. Manuals were often absent or assumed too much prior knowledge. More recently, teachers' guides have been included which describe experiments. The Papua New Guinea assorted science kit includes charts, reference books, slides, rock samples, a safety manual and mineral collections as well as equipment; it is really a multi-media package to support the other kits.

Lack of quality control is commonly reported as a major constraint to local production of kits. Quality is particularly threatened where teachers or students produce equipment. Early reports from SEPU in Kenya indicated up to 80 per cent loss in production of kits due to poor quality control. Handbooks have been prepared so that teachers and technicians can carry out routine tests on instruments provided in kits. The Regional Centre for Education in Science and Mathematics (RECSAM) in Malaysia provides training for the in-country testing and evaluation of prototype items.

It is not very meaningful to collate the average cost of science kits, since these vary considerably depending on their content. The cheapest kits can cost less than US$100. Thus, the general science kits for middle schools used in the Indian project noted above in Madhya Pradesh were being provided at a unit cost below US$50 with annual recurrent replenishment costs averaged between 5 and 10 cents per child in the mid-1980s. Kits for upper secondary schools in the same project averaged about 10 times more than this, reflecting the inclusion of more items related to the specialized teaching of separate science subjects. Science kits used in schools in Papua New Guinea in the late 1980s averaged about US$800-1,000 per school for each of five types of kit, all five being needed for a complete coverage of the curriculum. These kits are intended to supplement basic facilities rather than provide all that is needed to teach science. The kits were coupled with replenishment grants of US$700 per year per school.

The extent to which science kits may reduce foreign exchange costs depends on the proportion of local content. Some items cannot be produced where there is no industrial base or precision engineering. Typically, thermometers, electrical meters, accurate weighing devices, microscopes, and equipment that utilizes modern electronics and/or exotic materials is unavailable locally. On the positive side, there are increasing possibilities of using 'junk' material (UNESCO, 1985) as consumerism spreads and urbanization continues. Old automobile parts, bottles, discarded consumer electronics, oil drums and many other things may be available that can be used in science teaching. There is often resistance to this since adaptation requires imagination, time and some costs and may result in equipment viewed as second best, even if it is the best available. 'Street science' has possibilities but has to overcome prejudices in favour of conventional practical work (George and Glasgow, 1988).

In summary, kits do seem to have potential *advantages* in:

- reducing the costs of providing opportunities for some practical work and demonstration of scientific activities;
- possible reductions in foreign exchange costs;
- supporting local production capacity;
- encouraging the design of teacher- and student-friendly experimental activity that draws on material available locally;
- focusing thinking on the intellectual purposes of practical work.

For these advantages to be realized, a number of conditions need to be satisfied:

- science kits need to be integrated closely with curriculum learning objectives;

- the design of kits must recognize conditions in typical schools;
- teachers need some support in learning how to make the best use of material in kits and, where practical examining is required, this has to be feasible using the material that is provided, either in the form of a kit or otherwise;
- the replenishment of materials and repair of kit components must be simple, cheap and convenient;
- evaluation of the use of kits is needed to tune them to changing demands.

This first condition suggests that science curriculum development needs to be based on a realistic appraisal of the conditions which exist in the majority of schools. Specifically, experimental work has to be defined in ways which can be accomplished in the great majority of schools under the conditions that exist. There is a considerable difference in conditions in Malaysia, where virtually all secondary schools have purpose-built laboratory space and the majority of laboratories are adequately equipped, and those in Kenya, where the number of laboratories available in 1990 was about 1,950 and the estimated need was for 5,400 (Obura, 1992). Curriculum design based on the assumption of a fully equipped laboratory would clearly disadvantage the majority of students who do not have access to this environment. Here, practical activities should be designed that can be completed with materials in low-cost kits in normal school environments.

In most of the cases of the development of science kits which were reviewed, *training and in-service* support for the use of the kits was either non-existent or confined to an introductory period and not subsequently repeated. Generations of teachers after those who first used the kits, therefore, often did not have the opportunity to be introduced to the range of ways they might be deployed. Introductory exposure may have been insufficient for under-qualified teachers to appreciate the full potential of the kits. Typically it seems that in-service support for science kits has been associated with their introduction rather than as part of an on-going programme of support. In many cases, the impression persists that science kits are 'second best' and do not enable 'real science' to be taught, a view reinforced if in-service support for their use is not sustained.

Replenishment and repair is perhaps the weakest aspect of the production and use of science kits. In most cases, no systematic replenishment system, apart from periodic replacement, exists. This despite the obvious need to replace consumable material, repair equipment that breaks down and replace that which is worn out. Replenishment kits that include those components likely to be depleted each year, and support for replacement and repair facilities, are conspicuous by their absence. Rarely is a monitoring system in evidence that can identify where kits are in use, when they need replenishment and replacement, and how they might be improved to respond to emerging needs.

This last point made regarding the use of science kits, brings attention back to the purpose of providing such kits. If this is to promote a greater level of practical work in school science, it must be clear what the intellectual purpose of this is, and how activities relate to curriculum objectives for science. Paying attention to the definition of these purposes and the learning outcomes which are considered essential and desirable helps to clarify what needs to be provided. It might also result in a reappraisal of the need to provide very high-cost school laboratory facilities if other ways can be found to reach particular learning objectives for science.

Section 3: The provision of learning materials

Textbooks and other print material will remain the cheapest methods of providing learning resources to the majority of science students. There are a number of issues in the design, distribution and use of curriculum materials that have planning implications.

1. Some curricula dilemmas

Most countries have now experienced more than two decades of science curriculum development. In many developing countries (e.g. Morocco, Malaysia, Thailand), locally designed materials have replaced those imported from abroad, relevant material derived from national contexts has been substituted for that embedded in other cultures, opportunities have been taken to update topics to reflect recent advances in science and technology, and a wide range of different approaches to organizing the science curriculum have been adopted.

Most national systems now have a systematic curriculum development capacity in science, either through curriculum development centres directly, or through systems of commissioning and appraising learning materials produced commercially within a framework for approving recognized school texts. The characteristics of the curriculum development process within countries reflect different underlying philosophies on the centralization or otherwise of decision making on the curriculum and the provision of learning materials. In some cases, the same officially produced science texts are produced commercially and provided at some regulated price to all children (e.g. Morocco); in others, several versions are approved and commercially produced and local educational authorities decide which to use (e.g. Thailand). There are also countries where choice of science teaching material is made at the school level and many differing materials compete for the attention of teachers and students (e.g. Chile).

The most crucial and pressing issues relating to learning materials include:

- Is there an urgent need for the development of new materials as opposed to other interventions?
- How can material based on new insights into science learning, especially those that are culturally based, be developed and disseminated?
- What are the implications of the extensive development and use of 'workbook' or 'tuition' materials (including old examination papers)?
- How can continuing shortages of other learning materials be overcome – science print resources in local languages in the form of posters, charts, film/video clips, material to support out-of-school and informal-sector learning?
- In which ways can the development, availability and use of teachers' guides be improved?

2. Materials availability and patterns of use

In broad terms, three types of situation as to the availability and use of science curriculum materials appear to exist. In the *first*, curriculum development in science is in the third or fourth cycle since the 1960s: text material is widely available, quality is acceptable and relevance has been assured by extensive local development. In these systems there is often a choice of science curriculum material available to teachers and students. The problems of science teaching and learning are most likely to revolve around the factors that determine patterns of use of existing materials. It is then necessary to understand why the materials are not used. Reviewing and revising them, and training teachers on how to use them, may then be of greater urgency than developing new materials. Morocco is a case in point. The country has made extensive efforts to revise science curricula and related materials over recent years; however, many science teachers express reservations concerning their effectiveness and do not appear to use them extensively.

Development of new materials in cases similar to this is unlikely to be a priority unless and until reasons for lack of effective implementation are understood. The exception is where new materials are needed to satisfy changing definitions of science learning objectives. The most likely areas in which this may be true are where a greater emphasis is to be placed on technological aspects of science, and where environmental and/or science and society issues are to be introduced as a central element in the teaching of science. The other circumstance that may justify the production of new materials is where biases in presentation exist that are no longer thought appropriate or acceptable, e.g. biases of gender or ethnicity.

In the *second* type of situation, curriculum development activity in science may have not progressed beyond a first or second cycle. Courses designed during the 1960s and 1970s, derived from materials first produced in developed countries, may still be in use. Their relevance and topicality may be questionable. The availability and range of science curriculum materials may also be very limited, especially with respect to enrichment material, library books and relevant audio-visual resources. In many cases there may be an insufficient number of textbooks to provide for the needs of all children in secondary schools. The quality, relevance and quantity of science curriculum material may all be such as to suggest that the development of new curriculum materials is needed. The reasons why the curriculum development process has not responded to emerging needs may need careful consideration if attempts to develop new materials are to be successful.

In the *third* situation, provision of basic text material in the majority of schools is problematic. Purpose-designed science curriculum material related to the current syllabus either has not been developed, or has not been produced and distributed in appropriate quantities. What is available is dated, of questionable relevance and accessibility, designed for different contexts, and in poor physical condition. Where curriculum material does exist it may be of low quality, factually incorrect, and poorly structured. Significant amounts of material may be in languages not used by the majority of children and teachers. Acute shortages of print materials are likely to exist in most schools. National curriculum development capacity is likely to be weak and may be unable to design and develop curriculum materials of appropriate quality. Many sub-Saharan developing countries are facing this type of situation. Under these conditions, materials design and development are a priority. This may be a necessary but insufficient condition to improve access to and achievement in science. However, chronic problems of under-funding of education may have to be addressed before materials design, production and distribution can be successfully promoted.

168

In all circumstances, it must be remembered that the development of new curriculum materials is an expensive and time-consuming exercise. If quality is to be ensured and implementation effective, considerable investment may have to take place in the development and trial process and in supporting in-service activity to promote effective adoption. The decision should therefore not be taken lightly and a clear indication of the reasons why existing materials are inadequate is a precondition. If the problems lie more with the use made of materials, or their distribution, producing new materials may simply distract attention from the real problems of production, dissemination and support for effective use.

3. Insights into learning and curriculum development

The literature on how children learn science and acquire science constructs has expanded rapidly over the last two decades. A large number of projects, mostly in developed countries, have explored children's cognitive development, on the one hand, and the conceptual-izations they bring to science on the other (Wilson, 1981; Driver, 1983; Adey and Shayer, 1994). Despite the volume of this work, it is striking that in many developing countries little attention has been paid to the ways in which children learn science and what 'alternative frameworks' they may bring to the study of the subject. The concept of a zone of proximal development – "the distance between the actual developmental level as determined by independent problem solving and the level of potential development as determined through problem solving under adult guidance or in collaboration with more capable peers" (Vygotsky, 1978, cited in Adey and Shayer, 1994) has been widely recognized as an essential element in devising more effective methods of teaching science. However, few science curriculum materials have been constructed with the benefit of insights derived from this body of research on children's conceptualizations and capabilities. Even fewer practicing teachers utilize this as a basis for lesson planning. Child-centred approaches hardly penetrate actual science teaching and learning.

Hence, where the need to develop new science curriculum materials is justified because existing materials are seriously flawed, and adequate human and physical resources exist to undertake the task – new materials should accommodate structures derived from research on children's learning. It could be argued that there may be a conflict between the underlying philosophy of the 'constructivist' approach to learning and teaching and the development of written materials, especially those in textbook form: constructivist approaches recognize diversity in concep-tions and the personal nature of meanings attached to knowledge of the

169

physical world by children. It follows that a single text cannot hope to tackle the range of variation that may exist between individuals. This may be true, but it has the feel of a 'counsel of perfection'. In many countries, especially those with a limited supply of learning material in appropriate local languages, textbooks will continue to be central to the teaching and learning process. It would seem essential that their design takes cognizance as much as possible of what is known of children's conceptual structures, even if this cannot be comprehensively achieved.

4.   Workbooks and tuition materials

In most of the countries where print materials are widely available, parallel markets exist alongside those for officially recognized science curriculum materials. These markets produce and distribute print materials variously described as workbooks, key-facts material, tuition guides, etc., which are designed to capture elements of the science curriculum that are most likely to be examined and provide students with a minimum learning agenda intended to maximize the probability of passing public examinations.

Any discussion of curriculum materials development that does not consider these parallel markets is incomplete. Where the use of these secondary materials is extensive, it is tempting to conclude that the de facto curriculum is much closer to what they contain than the content and skills defined by official syllabi and approved curriculum materials. In Malaysia, for example, the case-study research established that the use of workbooks was very widespread for some courses, although these texts had been developed commercially by authors with no formal stake in the official curriculum and its goals. The emphasis of these materials reflected that of items in public examinations, which cover only a small proportion of the learning goals identified for secondary science.

It is unrealistic to attempt to discourage the production of these kinds of materials. Neither is it clear how this could be achieved, except perhaps where the market for text materials is heavily controlled. A service is performed if children who would not otherwise have access to printed material at least have access to privately produced secondary materials. Three observations seem appropriate. *First*, if official curriculum materials are of a quality judged equally effective in preparing students for examinations (and they are price competitive) they will be purchased in preference. *Second*, there may be a case to recognize some workbook materials as supplements to main texts if they are of appropriate quality and sympathetic to curricula goals. *Third*, since, in the circumstances described, it is often the content and form of public examinations that determine which learning is of value, the implication is that changes in the examining system may also influence the producers of secondary materials in ways which promote desired learning objectives.

5. Supplementary materials

Additional learning materials for science are in short supply in many countries. In some, the shortage simply reflects the absence of even modest funding for non-salary expenditures on science. In others, including many middle-income countries, problems are more concerned with the absence of local suppliers, the lack of material in national languages, and the low priority accorded to purchasing supplementary learning materials in many schools. As a result, many classrooms and science laboratories are largely lacking in interesting and informative charts, tables, photographs and scientific displays. Library holdings for science may be composed of different books on arbitrary topics, foreign language materials that are never borrowed by staff and students, and material with reading levels unmatched to those of the majority of students. Where these conditions apply, this represents a missed opportunity for quality improvement.

6. Teachers' guides

Teachers' guides provide a low-cost basis for in-service support for the teaching of science. They have the advantage that they are relatively cheap to provide to individual teachers and are durable, with a lifetime of several years. Teachers' guides are unlike in-service training courses, which are usually short, of variable quality, and which may not provide a full coverage of a science course. It would therefore seem that there need to be strong reasons to justify not producing teachers' guides in addition to syllabus material and student texts.

There is evidence that some teachers make limited use of teachers' guides. Where this is the case (as in Morocco, discussed earlier) the question is whether this is because the guide is poorly designed, or for other reasons. If the underlying cause is a lack of motivation and engagement by teachers with teaching, this has to be addressed. In any case, it would seem that developing teachers' guide material and ensuring its widespread availability is a necessary, if insufficient step towards improving teaching quality.

Section 4: Management of school science departments and support for science-teacher development

Systems for managing the implementation of the science curriculum in secondary schools vary from country to country. Typically, these systems have several elements that include the arrangements for managing resources at school level (teacher deployment, space use, utilization of support staff, curriculum quality control), staff development support for science teachers (in-service and on-service programmes), and school advisory and inspection systems.

At the school level, the development of science departments may be constrained by some or all of the following:

First, it is the exception rather than the rule in most countries that school principals have training and expertise in science (see *Chapter II*).

Second, it is also the exception rather than the rule that teachers who come to occupy senior science teacher posts have been trained for this role. Many countries do not have such a post as teachers resist any kind of peer control (as in France and a number of Francophone countries). Where it does exist, promotion based on experience rather than performance, leads to promotion to posts with responsibilities over science departments. Systematic preparation or induction is absent. Heads of science departments may thus acquire responsibilities for staff development, curriculum management and purchasing and stock control with little assistance in acquiring relevant skills. In such cases, their effectiveness as a manager tends to be limited.

Third, where administrative and financial power is vested in the principal, heads of science departments may have little influence over science colleagues, since their formal positional power may be very weak. It is characteristic of school systems in a number of countries that the school management structure is relatively flat, with a school principal and possibly a deputy taking responsibility for most decision making. The problems this creates can be exacerbated in situations where heads of science or science teachers with special responsibilities (e.g. head of science laboratory) may be junior to other staff in their own departments – for example, where recent entrants to teaching tend to be graduates and are promoted over older certificated teachers. It may also result in inefficient purchasing decisions if equipment and consumable items are not ordered by science-qualified staff, but by those without a science background in superordinate positions in the school hierarchy.

Fourth, successful utilization of expensive facilities for practical work depends on a concern for the efficient use of resources and a clear responsibility for monitoring and improving patterns of use. It is often unclear where responsibilities lie for maximizing laboratory space utilization, minimizing the use of consumable material consistent with effective teaching, and making best use of ancillary staff time.

Fifth, successful science teaching usually depends on clear objectives shared across subjects, and consistent laboratory practices designed to promote valued outcomes, over a substantial period of time. It is unlikely that these will be achieved unless there is co-ordination between teachers in approaches to practical work and common conventions applied to how it is written up and integrated with theoretical instruction. Effective linkages with other subjects, in particular between the three main sciences and with mathematics, are important. Language, as the basis of most school learning, also needs consideration. The question is where within school structures is this need for curriculum co-ordination considered and how is it achieved?

Our research suggests several options for addressing the problems identified. Leadership in science education and its management needs to be clearly located with staff who possess appropriate competencies. If most principals have little science background three strategies suggest themselves. First, where science departments and posts with special responsibilities for science education management do not exist, they need to be created. This may also help increase the motivation to become, and remain, a science teacher. *Second*, staff development and in-service programmes should incorporate modules specifically related to the management of science departments in view of their costs and curriculum significance. *Third*, delegation of authority over science department development should be encouraged so that those with specialist knowledge and technical skills have responsibility for achieving agreed objectives.

Science heads of department are likely to need bespoke training and support if they are to play greater roles in the management of their departments. As the science curriculum becomes more complex to manage it is critical that scarce resources are deployed efficiently. Promotion to this role should probably be associated with systematic in-service support, and appropriate reference materials on management practices should be made available. Groups of science heads of department should be encouraged to form self-help clusters so that the more experienced can share their wisdom with new appointees.

Heads of science departments should be given special responsibilities to ensure that scarce and expensive resources are used efficiently. This includes monitoring the use of consumable material and ensuring an appropriate pattern of re-supply; arranging maintenance and repair in a planned way, and making effective use of laboratory assistants where they exist. Grouping students to make good use of practical facilities through careful timetabling should also be undertaken. School laboratories can be significantly under-utilized in some systems.

Beyond the management of resources, the development of whole school curriculum policies on science are attractive and should be the responsibility of science heads of department. Where planning groups/senior management teams exist, their structures can be used for this purpose. Where they do not, ways of introducing them deserve to be considered.

Section 5: Maintenance issues and training of laboratory assistants

There was evidence in both the Moroccan and the Malaysian case studies of problems that can arise with the management of science equipment. In most of the schools studied the standards of equipment maintenance were not high and there appeared to be very little attention to systematic and preventive maintenance. In many cases it seemed that with sophisticated equipment the first fault was also the last, as it was not possible to repair the equipment locally. If this is the situation in relatively developed countries like Malaysia and Morocco, where the overall condition of science education is good, the situation is likely to be much worse elsewhere.

Laboratory assistants can help in ensuring maintenance provided they are well trained. Untrained laboratory assistants may be a potential hazard to themselves and to others unless restricted to the most basic tasks. Handling corrosive and toxic substances requires knowledge of proper procedures, as does the disposal of micro-organisms which may be cultured. The value of laboratory assistants lies in their ability to extend the utilization of trained science-teacher time, by taking care of the more routine aspects of lesson preparation, cleaning up, and maintenance and repair of equipment. Laboratory assistants are not normally involved directly in teaching students, although there may be some circumstances where they are available to help with different types of practical activity. Efficient laboratory assistance may therefore save several hours of trained-teacher time daily if productively employed.

Section 6: Patterns of in-service/on-service training of science teachers

Several patterns of in-service support for science education exist. The variety of approaches includes:

- in-service days where teachers are gathered locally at special centres or in designated schools to discuss particular topics;
- short in-service courses lasting up to a week, usually residential in regional centres run by national or regional-level staff;
- longer in-service courses lasting for three months or more usually associated with certification and upgrading of qualifications;
- school-based in-service support (otherwise known as on-service) during and after school hours, located in schools.

Very little appears to be known about the costs and effectiveness of these different approaches to in-service training and support. Where evaluations are conducted they are usually contemporaneous with the in-service events. They are very rarely undertaken after six months or a year to establish whether any impact on practice has been sustained. Independent evaluations by those uninvolved in the training process also appear rare.

Avalos (1995) has reviewed different approaches to in-service training and distinguishes between short and long courses, in-service through collaborative research projects and induction programmes for novice teachers. Her analysis cites individual examples of both short- and long-term training that had a measurable impact on teacher attitudes and practice. Research based in-service, usually employing the methods of action research, seem to be successful in heightening awareness of practice and opportunities to improve pedagogic style (Avalos, 1995; Stuart, 1991).

The value of *mentor support* is increasingly realized for the development of novice teachers, since this can provide structured developmental experiences during the first few years of professional development. In Japan, new teachers are required to partake in on-the-job training for at least 60 days a year in school under the supervision of an experienced teacher, and attend 30 days of day-release training (Iwasaki, 1991). Other studies (reviewed in Lewin, 1992) paint a very mixed picture, reflecting the heterogeneity of practice and the paucity of robust, and long-term evaluative data.

The most common reasons for arranging in-service programmes are those associated with the introduction of new science curriculum materials or the up-grading of under-qualified teachers. Typically the dissemination of a new course is planned to include exposure courses or longer training experiences to familiarize teachers with new learning objectives, content and pedagogy. Thus, in Thailand, introductory seminars are held for teachers whenever a curriculum change is introduced and a correspondence support service is provided to answer subsequent queries from teachers (Wilailak, 1991). 'Cascade' models are quite common where national workshops are followed by progressively more local-level activity organized by those who have been on national courses. Up-grading programmes are often institutionally based in teacher-training facilities, taught by staff also occupied in initial teacher training, and tend to be relatively lengthy. Korea, for example, operates all of these patterns and additionally requires science teachers to attend 'refresher' courses every three to five years (Lee, 1991).

Evaluation of in-service activities is problematic for the reasons given above (see also *Chapter II*). Which types of in-service arrangements are most effective for different purposes will depend on a range of factors, some of which will be country specific. There are grounds for concern that so little information is currently available on the comparative advantages of different approaches to training and the extent to which they result in sustained changes in teaching and learning practices.

More emphasis on school-based in-service and on-service support may be attractive in some systems, where sufficient trained staff exist to organize activities at school or local-cluster level. These have the potential to provide support and advice as and when it is needed, to allow the continuity of assistance that is generally not available through sporadic short courses, and to encourage collaborative approaches to the development of teaching and learning at the school level. They are also likely to be cheaper to deliver. To be successful, this kind of approach benefits from the provision of written materials for teachers, around which in-service/on-service development activity can be based.

Costs clearly do vary considerably between the different types of delivery. Systematic analysis of cost options, especially linked to measures of effectiveness, simply do not seem to exist. This is curious since in-service courses are frequently requested by teachers, and represent a major budgetary item, especially when curriculum change takes place. Residential courses will be much more expensive than non-residential local courses. On-service support foregoes little teacher time, whereas full-time courses have a substantial cost in lost teaching time if run during the school year. Several in-service days spread over a period may be more effective than all clustered together, depending on the purposes and method of training, though they may be more expensive.

The cheapest forms of in-service support are almost certainly those that are based on printed materials and which are delivered locally or through school-based sessions run by local animateurs.

Section 7: Inspection and advisory systems

Structures that exist above the level of the schools to monitor and support science education generally include school inspection systems and, less frequently, subject advisers. The latter may or may not be delegated some of the functions of inspectors. There are several issues raised by current patterns of inspection and advisory systems to support science education.

First, there may be ambiguity between the inspectorial and advisory function. Often, the former may be dominant and inspection systems may focus on bureaucratic indicators of performance (attendance, inventories, staffing numbers) at the expense of providing advice and assistance to overcome problems identified.

Second, inspectors may not have appropriate powers to correct deficiencies that they identify in physical provision or in the posting of science-qualified staff.

Third, the coverage of schools may be problematic. For example, two science curriculum advisers are provided for each state in Malaysia, independent of size. In the larger states, it is physically impossible for these advisers to visit more than a small proportion of schools. In Morocco, the ratio of science inspectors/secondary schools is more favourable, but the existing capacity of teacher support through science inspectors remains nevertheless very limited.

Fourth, practices vary in terms of which schools receive most visits from inspectors and advisers. Two categories of schools, those generally regarded as the best, and a selection of those with very poor results, appear to attract more interest than others according to the Malaysian data. This means that under-performing schools which nevertheless may achieve average results are overlooked. Low-scoring schools are put under a general review and attempts made to monitor and support measures designed to promote improvement.

Fifth, detailed and up-to-date data on school performance, staffing and resource utilization are not always available. This is likely to limit the effectiveness of the inspection and monitoring process.

Conclusion

This chapter investigated different cost-effective approaches to science education provision. Specialized science schools appear an attractive option for those countries which need to develop a stock of highly qualified science-trained cadres, but do not have enough resources to provide adequately for all secondary schools in the country. One way or another, all countries have prestigious institutions which play a similar role. When the countries' level of resources – human and financial – increase, the justification for highly expensive and selective schools tends to diminish. The disadvantages – such as the risk of wrongly selecting students and of leaving out some potentially good future scientists, or of discouraging non-selected students in other schools – increase and may outweigh the advantages. Kits and multiple-purpose science rooms are an option worth considering as an alternative to very expensive laboratories. However, equipment and consumables need to be replaced and kits have to be regularly refurnished. To be worthwhile, investments in practical facilities should be accompanied by efforts to increase teachers' motivation as well as their skills in planning flexible and creative lessons, support should be provided to science teachers at school level through the production of teaching guides, the distribution of enrichment materials and other learning materials, the organization of school-based training sessions run by animators, and senior science teachers, inducted to play a role in curriculum co-ordination and to lead science-teaching teams, should receive systematic training. Finally, textbooks and other print materials are and will remain the cheapest method of providing learning resources to the majority of science students. The production, distribution and use of relevant material remain priorities in many countries.

Chapter V
Information base on science education

The research work undertaken for this project, both in the framework of the survey and for the in-depth case studies, indicates that data on science education are not easily available in some countries. To obtain information for this study it was necessary to collate data from varying sources in and outside the Ministry of Education, and to conduct surveys and other data-collection exercises. Data were scarce concerning the coverage of science education and the take-up of science, i.e. how many students study science at different levels and differences in the degree of participation by gender, regions, social groups; the specific teaching conditions of science; the level of pupils' achievements in science; the flow of students into higher education, further studies and the labour market; and last but not least, the cost of science provision. In this chapter, a number of selected indicators and types of analysis are listed which seem pertinent to help the planner and the decision-maker monitor the status of science education and its performance (*Section 1*).

The project involved a fairly wide range of data collection techniques. In *Section 2*, we comment on the approaches used.

Section 1: Information base on science education

There are several approaches that can be used to establish the extent of participation in science education, the quality of what is provided, and its cost. Most of these have been employed in the course of the IIEP research programme. First, existing data can be analyzed; the usual sources are to be found in school census data (most countries have an annual exercise in which every school completes a questionnaire), records of capitation and other expenditure, and examination entries and performance data. In addition, special studies may have been undertaken by ministries or universities and research institutes which relate to science education.

Where there are gaps in this secondary data it may be necessary to conduct a survey on a national sample to probe into the reality of science education at different levels. Special surveys can examine questions not normally covered in the school census, e.g. transition rates within schools into science streams, actual patterns of option choice, the proportion of qualified teachers teaching science and the proportions of qualified science teachers teaching other subjects at different levels, the extent of in-service support for science teachers, the adequacy of physical facilities, etc.

A third method of data collection is to undertake detailed school case studies which involve field workers spending substantial periods of time in selected schools. This is often the only way to get insight into school-level allocation of resources and the underlying reasons for patterns of teaching and learning. Studies of this kind allow qualitative data to be captured.

Fourth, assessment data are usually capable of analysis beyond that conventionally undertaken by examination authorities. They can provide the basis for studies which identify relatively effective and ineffective school science provision in terms of achievement outcomes. Where such data are based on instruments that do not collect data appropriate to a specific purpose (norm-referenced tests may not be designed to identify underlying learning difficulties, not all students may take public examinations, scientific literacy may not be assessed, etc.), it may be desirable to design monitoring tests which can sample a wider range of pupil performance. These can be applied selectively on a sampling basis to limit the logistical costs of the exercise.

For many planning and policy questions the most cost-effective way to collect and analyze data on science education issues is likely to be to add a number of questions to the annual Ministry of Education question-naire and/or to carry out surveys on samples of schools. It may be useful to design targeted research studies that use some or all of the methods of data collection mentioned above. These may go beyond the condensed case studies typically used for overview exercises and probe more deeply into aspects of learning and teaching methods. These may uncover the characteristics of learners, teachers and school and system management which are relevant to planning improved provision.

1. Coverage and participation in science education

A starting point for an analysis of coverage and participation in science education is the construction of a flow chart of students' enrolment from primary through secondary schools into further and higher education, separated into curricula options. This can illustrate the proportions of those enrolled at each level who are experiencing some kind of science education. *Table 5.1* below was created in the Malaysian baseline study to provide such an overview.

Table 5.1   Malaysia: Students taking science as a percentage of total enrolment and population by levels of education (1989)

Level of education	Type of science curriculum	Enrolment in science	Total enrolment	Total pop.	Percentage enrolment in science
		(a)	(b)	(c)	× 100
(a) Primary (6 + to 11 + year)	Man and the environment	2390920	2390920	2415400	100.0
(b) Lower secondary (12 + to 14 + years)	Integrated science (KBSM)	938518	938518	1132600	100.0
(c) Upper secondary (15 + to 16 + years)	General science Pure science and science/technical-based subjects	252365 106243	358608	730500	70.4 29.6
(d) Post-secondary (17 + to 18 + years)					
Form 6 Pre-University	Pure science Science-based courses	14427 } 9699 }	} } }	} } }	
Higher education/college		}	}	}	
Polytechnics	Science-based courses	7449 } }	} }	} }	
Teacher training	Science options	139 }	}	}	
Colleges		46400 }	121042 }	736800 }	38.3
ITM	Science-based courses	10888 }	} }	} }	
TAR college	Science-based courses	3798 } }	} }	} }	
(e) University (19 + to 24 + years)	Pure science and science-based courses	22460	53476	2007800	42.0

Source: Sharifah Maimunah and Lewin, 1993.

181

Where science students are separately streamed into specialized science classes constructing such a diagram should be relatively straightforward. Problems will arise with this approach when the number of possible subject combinations is large and when the definition of 'science student' is ambiguous.

More detailed analysis of coverage, and exploration of changes that may be taking place as a result of shifting patterns of student option choice, depend on adequate statistical information from school census data that can identify the numbers who are enrolled in science in each grade. This can be disaggregated in a number of different ways – by sex, type of school, location, level, science subject, and by ethnic or language group.

It will be of interest to know how many girls study science to what level. In many countries, efforts are being made to increase female participation and involvement in science. Differentiation in participation by sex is often associated with a specific transition (from lower to upper secondary or from upper secondary to pre-university) rather than being consistently distributed at all grade levels.

It may also be important to identify patterns of science enrolment in rural and urban schools or in schools which enrol students likely to proceed to tertiary institutions, and in schools where the majority of students enter the labour market at the end of the school cycle. The dynamics of participation may vary between these types of schools and aggregation can conceal patterns that may exist. It is therefore important to monitor access to science for those likely to occupy jobs with a substantial scientific component and for those entering the general workforce.

Patterns of option choice can determine the proportion of the cohort that experiences different types of science education. It is important to establish what range of choice is available and who makes the choice of option on what criteria. In some cases, students are allowed a free choice of combinations of subjects; in others, choice is heavily constrained. Allocation may be directive or permissive. In the former case, some qualified students opting for science may subsequently drop science specialization. If this is so, the reasons for it need to be explored.

It is relevant to try and establish enrolment ratios in science studies based on an analysis of the cohort of school-age children at a particular grade level. Simple proportions of enrolled students studying science can be misleading if enrolment rates vary greatly between levels. The ability to calculate overall participation rates depends on the availability of appropriate demographic data.

2.   Teaching conditions

Quantitative indicators that may be available from secondary data and have a direct bearing on the conditions under which science is taught include:

- pupil/teacher ratios for science (using both qualified and unqualified science teachers as the denominator and number of science and non-science students as the numerator);
- proportion of science teachers teaching science who are not science trained and proportion of science teachers who do not teach science;
- ratio of trained science teachers to the number of students studying science;
- average number of teaching periods of science teachers;
- proportion of untrained science teachers by subject taught;
- ratios of science support staff to science teachers and to the number of laboratory rooms;
- ratio of science inspectors to science teachers and heads of science departments to science teachers;
- ratios of laboratories to science students in general education and at levels (in institutions) where there is specialization;
- rate of utilization of laboratories and science rooms;
- proportion of students without all the necessary textbooks, and without any textbooks.

The appropriate ratio of science teachers to students is system-specific since it depends on the amount of curriculum time allocated for different types of science and the mix of students at different levels following different options within a school. Norms are not difficult to establish against which comparisons can be made. Variations in this ratio (by type of school, district, region), and in the ratio of qualified to unqualified science teachers, can indicate where provision is problematic.

The proportion of science teachers who are not trained as such is also important information. In Malaysia, about 70 per cent of the teachers teaching science were trained to do so. At the same time, about 17 per cent of science-trained teachers were not teaching science. There was a wide variation in this figure between states from 53 to 77 per cent (see *Table 5.2*). The ratio of trained science teachers to the number of students studying science ranges from 1:129 to 1:167 in Peninsular Malaysia and the ratios for Sabah and Sarawak were 1:300 and 1:242 respectively. An important issue to measure with an appropriate indicator, therefore, would be the distribution and deployment of science teachers.

Table 5.2   Malaysia. Status of teachers teaching science by state

	Teachers teaching science				
	Sc. Option (per cent of total teaching Sc.)	Non-science option (percent of total teaching Sc.)	Total	* Total no. of students studying science	No. of students per trained sc. Optionist
Johor	996 (72.2)	383 (27.8)	1 379	156 682 (11.8 per cent)	157
Kedah	593 (68.7)	258 (30.3)	851	96 835 (11.8 per cent)	163
Kelantan	590 (74.0)	207 (26.0)	797	84 531 (10.3 per cent)	143
Melaka	302 (74.2)	105 (25.8)	407	47 362 (5.8 per cent)	157
N. Sembilan	394 (70.0)	169 (30.0)	563	5 737 (7.3 per cent)	152
Pahang	459 (62.7)	273 (37.3)	732	76 482 (9.3 per cent)	167
Perak	1 041 (77.4)	416 (28.6)	1 457	173 625 (21.1 per cent)	167
Perlis	103 (76.9)	31 (23.1)	134	13 317 (1.6 per cent)	129
P. Pinang	547 (74.8)	184 (25.2)	731	86 106 (10.5 per cent)	157
Selangor	857 (75.1)	284 (24.9)	1 141	139 780 (17 per cent)	163
Terengganu	352 (76.0)	111 (24.0)	463	48 067 (5.8 per cent)	137
Wilayah Pers.	533 (76.0)	211 (28.4)	744	83 030 (10.1 per cent)	156
Sabah	331 (52.8)	296 (47.2)	627	99 442 (12.1 per cent)	300
Sarawak	489 (56.1)	382 (43.9)	871	118 187 (14.4 per cent)	242
Total	7 587 (69.6)	3 310 (30.4)	10 897	1 283 183 (100 per cent)	169

*   Includes all lower secondary and upper secondary students as well as sixth-form science stream students.

Source: Adapted from the Government of Malaysia's Human Resources Development Plan Project, Report on Education (1990), Economic Planning Unit, Prime Minister's Department.

How much laboratory provision is made available depends on the amount of time allocated to laboratory-based practical activity, bearing in mind that there may be a minimum provision needed to justify offering a particular science subject. Laboratory utilization rates can provide an indicator of whether provision is adequate, although these kinds of data are generally only available from school level surveys. Simple room occupation rates must be treated with some caution since it may be that laboratories are used for normal class teaching.

The number of students per laboratory in the Malaysian case-study schools varied from about 200:1 to over 350:1. What the ratio should be depends on the mix of classes between science and non-science and upper and lower school enrolments. Thus the ratio is not always easy to interpret. In Morocco, the average ratio of 351 pupils to a specialized room in lower secondary schools and 176 pupils in upper secondary schools illustrated among other things the fact that upper secondary schools were relatively over-equipped while lower secondary schools were under-equipped. This was confirmed by other indicators such as the proportion of lower secondary schools with no specialized science rooms and the rate of utilization of specialized rooms in different types of schools.

Practice in relation to support staff and their distribution varies widely from no provision to generous levels that may exceed 25 per cent of the teaching staff. What is appropriate will depend on the extent to which the curriculum requires practical activity, the teachers' teaching practices, the equipment available in schools, and the resources allowing such staff to be recruited. At lower levels where simple practical work is expected, ancillary staff are not usually needed in substantial numbers. At higher levels, the demands may be greater and it may be more cost-effective to use ancillary staff than science teachers to maintain laboratories and equipment. Competent assistants should be able to take responsibility for maintenance of several practical laboratories. Information should be collected on their level of training and their deployment.

Where the ratio of printed curriculum material to students is low it is likely that there is poor quality provision. Provision has to reflect what can be sustained financially and what is essential for teaching the existing science curriculum. Estimates of the number of pupils with no textbooks or sharing textbooks can be obtained through a teacher questionnaire or through a student questionnaire using appropriate sampling. The causes of under-provision are likely to be related to the policy on purchase and distribution – e.g. whether books are provided free or have to be bought individually and whether distribution is a public or commercial responsibility.

Inventories of equipment may be available, indicating what items of equipment schools possess. Full inventories are very expensive to conduct and cumbersome to analyze. They also become very rapidly out of date. If carefully constructed they could provide a rough guide to the conditions under which practical work is taking place; thus they could help in monitoring the need for the purchase of equipment. The information collected should remain at local or regional level, to be used by the inspector to scrutinize equipment demand from schools. In designing inventory lists for the purpose of monitoring it is important to distinguish between items which are relatively expensive and those that are cheap; those which are used repeatedly and those which have a single use; and items for classwork and those for demonstration. One way of doing this is to analyze equipment demand in relation to curricula according to frequency of use, using a matrix of the kind shown below.

Item	Number of periods of planned use	Unit cost	Unit cost per student per period

This kind of inventory has only been undertaken in one of the countries included in the research (Morocco). To be useful, equipment lists should be drawn up on the basis of expert opinion of what is thought to be needed to teach the science subjects, keeping in mind the likely patterns of use and costs of purchasing different types of equipment.

3. Achievement data

Assessment data are critical to appraisal of the condition of science education in terms of output and demonstrated learning gains. Whatever the level of physical provision, and whatever the evidence on teaching and learning practices, unless there is some evidence that science constructs are being acquired and appropriate knowledge and skills can be applied, investment may be judged ineffective when learning gains are evident.

Aggregate performance data are usually available in systems which have national examinations. This provides a starting point. Overall pass rates based on individual and school-level data can give some indication of the learning benefits that arise from a particular cycle of schooling, assuming the test instruments used are valid and reliable. More sophisticated analysis is usually needed to provide sound policy advice. There are a number of reasons for this:

• Pass rates generally conceal actual levels of achievement if they are based on norm-referenced instruments – it is raw scores that illustrate what candidates can actually do.

186

- Aggregation (e.g. average school pass rates) may conceal very different patterns of performance within schools which have the same pass rates.
- Pass rates alone take no account of the 'value added' by the school.
- Dissatisfaction with 'below average' performance cannot lead to policy to bring all schools to above-average levels of performance.
- Intervention to improve performance necessitates realizing the nature of the different types of tasks that make poor candidates fail. This can only be achieved by analyzing performance at the individual item level.

As reported in *Chapter III*, the Malaysian study illustrates in detail some of the types of analysis that may be useful. When raw scores were explored it became clear that levels of achievement in science of significant numbers of students were unsatisfactory. Analysis of aggregate data indicated that substantial numbers were failing or scoring marginal passes which represent low levels of achievement. Within the group of schools studied it was clear that some schools achieved a given pass rate by maximizing the number that reached the minimum standard and achieved bare passes; others had a different distribution which placed more emphasis on high achievers so that substantial numbers of distinctions were scored and fewer bare passes. An example makes the point. Schools A and C in the example below have similar overall pass rates. However, school C has far more high-grade passes than school A. Although school B has the lowest overall pass rate, it has the highest proportion of high-grade passes. If a weighting system is used for aggregation (Distinction=3, Credit=2, Pass =1) the differences between schools are captured in a different way.

Grade	Distinction	Credit	Pass	Fail rate	Overall pass rate	Weighted score
School A	4	25	65	6	94	127
School B	15	35	35	15	85	150
School C	13	41	40	6	94	161

It remains the case that this type of aggregate data disguises differences that exist in the quality of the intake to the school. Relatively low pass rates in marginalized rural schools are probably more difficult to achieve than the 100 per cent pass recorded by some favoured urban schools. Thus apparently below-average schools might be performing more effectively than those with above-average pass rates. A simple way of indicating 'value added' is to moderate secondary school examination performance scores with a factor derived from the entry-level scores of students on the primary school Standard 6 assessment. This would show which schools were able to improve the relative levels of achievement of their students and which did not and would focus attention on under performing parts of the system rather than simply on below-average schools.

Standard techniques are available to analyze multiple choice questions. Thus, for example, facility values can be used to indicate the proportion of students who identified a correct response from a list of alternatives (those questions with the greatest facility values are the easiest and vice versa). Facility values can vary between one (all candidates get the correct answer) and zero (all candidates fail to identify the correct answer). Questions can then be arranged in order of difficulty.

Facility values for individual questions can be simply correlated. Thus, an examination question might have a facility value of 0.75 (75 per cent of students answer correctly) for rural students and 0.85 (85 per cent of students answer correctly) for urban students. This question would, therefore, be easier for the average urban student. A refinement of this is to consider the relative difficulty of different questions by asking:

"How does this question compare in difficulty with other questions for rural students? How does this relative level of difficulty compare with the relative difficulty of the same item for urban students?"

A question might be the fifth easiest for rural students and the twentieth easiest for urban students. It would therefore be relatively easier (compared to other questions) for average rural students despite probably being absolutely more difficult. This approach was adopted in some of the analyses designed to identify problem areas in the curriculum. More sophisticated methods can be adopted using discrimination indices, etc., if statistical expertise is available. Item-by-item analysis was conducted on a sample of about 5,000 lower and upper secondary students. This revealed a number of interesting patterns, including the following:

- Low-scoring students were not especially disadvantaged by higher cognitive-level questions; where these questions were appropriately contextualized they sometimes did as well as high-scoring students.

- Rural students were apparently disadvantaged by items that assumed specific knowledge and experience of laboratory-based science.

The significance of these kinds of findings for planning to improve performance have already been extensively discussed in *Chapter III*. This illustrates how important it is to be sure that what is measured is a function of the quality of the teaching rather than the quality of student achievement. It also draws attention to the value of developing indicators of school performance which recognize that the distribution of grades is as important as overall pass rates – which is preferable: all students reaching a minimum level or some achieving high grades? It reminds planners that some concept of 'value added' is important in judging school performance and effective practice. Finally, the methods offer the opportunity to feed information back to schools on those areas of the curriculum where students perform particularly poorly so that strategically targeted interventions can be designed.

Several of the countries studied do not have national systems of examination, or restrict examination to particular points in the secondary cycle. Prime facie in these countries there is a strong case to develop monitoring instruments at selected points in the secondary cycle to allow judgements to be made about the level and distribution of achievement that result from investment in science education. Without this it will be very difficult to establish where intervention is needed and where investment is wasted. Any attempt to identify effective schools and factors affecting achievement will flounder without this type of data.

Where assessment data are piecemeal and derived from non-standardized (and often unmoderated) school-level tests, there is also a need to design national monitoring and assessment systems that can provide objective evidence of standards of achievement in science education. The IEA science studies provide much informative and useful data on the construction of appropriate instruments. They also illustrate the dangers of crude cross-national comparisons which are of limited value operationally. Monitoring instruments should be contextually defined to relate to specific national curricula objectives and their analysis conducted in the light of these. Many countries are beginning to work on their own systems, (Ross, 1994a, 1994b; IIEP, 1995).

Decisions have to be made about the basis for national monitoring which recognize the costs and the analytical capacity available. Light sampling may be sufficient to develop order-of-magnitude insights into levels of science achievement in different types of schools and different regions. More comprehensive sampling is likely to be needed for detailed intervention. Where national examinations already exist these may provide the basis for monitoring achievement if detailed analysis can be undertaken along the lines indicated above. Though the need to establish a purpose-designed monitoring system is weaker in these cases, it should still be considered as an option, since norm-referenced national examinations designed to select are directed towards a specific and different set of objectives.

4. Destination of science school leavers in higher education and employment

The study of destinations of secondary school leavers may be analyzed in terms of those who proceed to higher levels of education and training. This involves the listing of all possible learning opportunities open to secondary science school leavers and the accumulation of statistics on the intake to post-school institutions.

There are two ways of obtaining information on this: one is to do a cross-sectional analysis, as was done in Malaysia; another way is to follow up students into higher education. The second approach is obviously more accurate. Morocco follows all of the secondary school leavers who enter higher studies in the country in a systematic way. This is part of the system of monitoring of students and scholarship holders. The information which is collected includes the following:

- name of the student;
- age;
- sex;
- average standardized score on the secondary school certificate ('Baccalauréat') and credit obtained;
- single-subject scores at the secondary school certificate;
- year of first enrolment in post-secondary education and field of study;
- examination results at the end of each subsequent year (pass or not);
- field of study and year of study for the following years.

With this information it is easy to produce tables on the number of students enrolled by field of study at the beginning and the end of each school year and reconstruct a cohort of students by field of study, year after year. Diagrams, such as the one presented in *Chapter II – Section 7* illustrate the flow of students through higher education in Morocco. Indicators such as: the overall transition ratio from secondary to post-secondary education, the transition ratio by field of study separating out science from non-science fields, and the proportion of secondary science leavers who obtain different diplomas and degrees after different numbers of years enrolled can be obtained. This information is very important for planners and decision-makers. It provides useful information to adjust the number of people admitted to different streams at secondary level and suggests appropriate admission policy at higher education level. A number of countries have similar information available, and could usefully do this kind of analysis.

Data on the absorption of qualified science school leavers into the labour market are generally much more difficult to obtain than data on flows into further and higher education. This requires labour force census data and surveys of employers which may be beyond the scope of educational planning divisions to undertake. Proxy measures can be illuminating. It is, for example, relatively easy to analyze vacancies advertised in the local press to establish which categories of school leavers are in most demand and what conditions (in terms of examination passes) employers place on entry to jobs. This may indicate scarcity or surplus; it may also indicate what trends are occurring. This can be coupled with projections that may exist on the likely growth of the workforce, subdivided by occupational category and level of qualification anticipated. Although these will never provide precise indications of how many science-qualified school leavers may be needed in future years, gross imbalances are likely to be exposed.

The study of destinations addresses a critical planning question. If it is not known what happens to graduates specialized in science at school level how can judgements be made as to whether there are too many or too few? A competent tracer study would be the best solution and there is a strong case to conduct these periodically, if only on a sampling basis, in every country.

5. Costs

None of the countries included in the IIEP study have systematic and recurrent monitoring systems to keep under review the costs, effectiveness and needs for science education. Information on each area is piecemeal, inconsistent and sometimes entirely lacking.

Expenditure per student on science is a difficult indicator to use. Capitation rates – where they exist – may give some indication of what is nominally available, but do not usually indicate what the money is spent on or if most is spent on equipment and consumables for students in the higher grades and little on lower-school science. Most flat-rate capitation systems penalize small schools since individual items of expenditure may represent a larger proportion of the total available. They may also favour older students at the expense of younger ones since, typically, more expenditure is devoted to equipment for use at the higher levels, where enrolments in science are smallest. Detailed insights are only available on actual expenditure patterns through school level research. A specific survey may be required to analyze school budgets.

Section 2:   Approaches to data collection and analysis for planners

The methods used for data collection in the IIEP case studies had five elements: a baseline study, a national survey, case studies of selected schools, an analysis of assessment data and a review of the destinations of secondary science graduates. We have already commented on the last two elements and offer observations on the value of the other methods below.

The purpose of the baseline studies was to draw together secondary data on the flows of students, enrolment rates, provision of teachers, curricula patterns, and resources available. Data were accumulated from the school census, annual statistical reports, small-scale enquiries directed at private secondary school institutions on which there are no national data, and various records located in different parts of the Ministry of Education. Interviews were also conducted with appropriate staff to help identify what issues were considered most problematic. Baseline studies have considerable value where sufficient secondary data exist to support their creation. Purposeful analysis of available data can draw attention to unknown or unappreciated aspects of the way the science education system is functioning which are not obvious when data are discretely organized in separate places for many different purposes. In both Morocco and Malaysia, the baseline studies allowed the identification of key issues that were to be followed up in subsequent analyses.

The next step was to consider the need for school-based case studies and special surveys. Surveys are a cost-effective way of establishing the level of qualification of teachers, their utilization patterns, how widely key pieces of equipment are distributed, whether adequate space is available according to explicit criteria, the distribution of ancillary staff, etc. Case studies provide more insight into actual patterns of learning and teaching and equipment utilization than do surveys. They may be at least as informative in suggesting policy options that would have an effect on practice, based on the qualitative insights arising from observation and interviewing of science teachers.

In both Morocco and Malaysia, a conventional national survey was conducted. In Morocco, a probability sample of all secondary lower and upper secondary schools was drawn, while in Malaysia, for logistical reasons a sample of schools was taken from four states only which were broadly representative of the levels of socio-economic development in the country. In both case studies, two types of questionnaire were employed – for principals, and for all science teachers in the selected schools.

The questionnaires used were designed in a residential workshop after preliminary investigations by the research team in a small number of schools, and were pre-tested and reviewed. These were administered using district officers to personally deliver them. In Morocco, questionnaires were also personally collected. This ensured a high response rate (100 per cent) and the checking of information concerning school data (in the principals' questionnaires). In Malaysia this type of data collection and the personal checking of the accuracy of the data was not possible for logistical reasons. The response rate was therefore lower, and so was the reliability of some of the information obtained. The principals' questionnaires included biographical data, information on staffing and the organization of teaching, patterns of enrolment in science, levels of achievement, and opinions about aspects of science provision. Teachers' questionnaires focused on individual science teachers' characteristics, on the description of their teaching practices during a single month, their perceptions of teaching and learning, and the adequacy of resources.

Analysis of the survey data used standard statistical techniques and the usual normal procedures for cleaning data, consistency checks, preliminary analysis, aggregation of appropriate data categories, etc. The analysis of each survey instrument was written up separately first and then incorporated into a synthetic report.

The case-study research used methods that may be less familiar to planners. It is therefore described here in a little more detail. Eleven state schools were identified in the four states selected for case studies in Malaysia. Two special science schools were added to this sample as they were clearly of particular interest. No selection of schools could have been representative of the survey sample of schools since the sampling ratio was too small. About one person-month was allocated to each case-study school and this limited the total number of case studies that could be completed in the appropriate time frame.

Various data collection techniques were employed to ensure that data generated from the case studies both captured the special characteristics of science education in the schools and also covered a common agenda of issues on which insight was needed. The methods used ranged from those which were relatively unstructured and responded to issues raised by respondents, and those which were more structured and reflected matters of general concern identified from other parts of the research programme. Corroboration across the data was sought wherever possible to cross-check important observations. The initial agenda for the case studies was identified from field visits to four schools around Kuala Lumpur, including those with a rural location, discussions with appropriate professional officers of the Ministry, and documentary analysis of reports and evaluation studies. Preliminary case-study instruments were developed to collect specific information, e.g. on consumable materials, equipment, allocation and expenditure of funds for science, physical facilities and resources, and teaching and learning materials.

Drafts of case-study instruments were piloted in two urban schools, amended and tried out again in a rural school. These instruments were finalized at a residential workshop. The main structured instruments used in the case studies are indicated in *Table 5.3* below, along with an indication of other sources of data.

The case studies were conducted in two phases (the detailed case study and a follow-up visit). Planning was done to ensure that fieldwork did not coincide with examination periods or other events that disrupted the normal working of the schools. Scheduling also had to make allowances for the different weekends observed in the different states. The relevant education district offices and case-study schools were invited to participate and given information on the duration and nature of the activities. A co-ordination meeting preceded each fieldwork exercise where case-study researchers agreed on a division of responsibilities and timetabled their activities. In all cases, teams met with the school principal and/or senior assistant first and negotiated the pattern of work. Courtesy visits were also made to the District Education Offices and the State Education Department where this was possible.

The first visits lasted five to six days in each school. In the course of this, baseline data on the school and its staff and students were acquired, relationships with staff were built and assessments made of the facilities and equipment available. The conditions of teaching and learning science were also explored through interviews with teachers and observations of some lessons. Follow-up visits were made after preliminary analysis and writing up. During the follow-up visits, information gaps which had been identified were filled and a further explanation of problems and issues was explored. With four teams of researchers working in pairs, the fieldwork and the write-up of individual case-study reports was accomplished in about 14 weeks. Requests for detailed information, e.g. on income expenditure and examination results, were made early in the visit to enable school records to be consulted and information to be compiled. Checklists were used to standardize some data collection. Day-to-day adjustments were made on the basis of evening co-ordination meetings. Case studies in the same states were undertaken concurrently, allowing inter-team discussions to be carried out.

Table 5.3   Case-study instruments and other sources of data

Type	Focus
Interview Schedules	• Principal • Senior Assistant • Science Teacher • Laboratory and Technician • Guidance and Counselling Teacher • Library Teacher
Observation	• Lessons • School environment/ Climate
Checklists	• Laboratory equipment • Consumables • Furniture/Fittings • Science funds • Physical facilities
Timetables collected	• Laboratory • Teachers/Personnel • Class
Documents examined	• Visitors' book • Student exercise/Activity • Books • School blueprint • Minutes of departmental meetings • Circulars • School magazines • Examination (SRP/SPM/STPM) Results • Textbooks

Semi-structured interview schedules were prepared for each category of informant. This ensured that key issues were not overlooked. The format also allowed interviewees to raise issues of particular concern to them. Thus, researchers were not required to use interview schedules rigidly so that interviews could flow as naturally as possible. Interviews with school principals/senior assistants were aimed at establishing basic information about the school, the pattern of provision of science, and the problems associated with this. Interviews included personal biodata, school-level data on the catchment area, facilities, staffing, timetabling, finance, perceptions of the aims and problems of science education, effectiveness of the support system and the destinations of school leavers. Science-teacher interviews delved into areas which included biodata, teaching loads, personal aims of teaching science, teaching conditions, patterns of teaching and assessment, problems of teaching science (including the identification of difficult topics), the value of in-service training, and provided insights into teachers' motivation and commitment. Interviews with laboratory technicians encompassed areas such as biodata, workload, resources available as well as perceptions of teaching and learning problems.

All available science-related school documents were studied. These included the school blueprint, which spelled out priority areas for improvement at the school level. The visitor's book was checked, as this provided a record of all visitors to the school. It was thus useful in determining the extent and frequency of support and advisory visits related to science. Records of income and expenditure provided information on the pattern of expenditure on science. Students' workbooks and exercise books were analyzed to determine the type and frequency of assignments and practicals and the general quality of work. Class and teachers' personal timetables were used to fix appointments for teacher interviews and classroom observations as well as to analyze the distribution of science periods. Laboratory timetables were analyzed to ascertain the rate, frequency and distribution of laboratory use.

The writing up of case-study reports took place after researchers returned from the field. There was no standard division of the responsibility for writing up the case studies and each research team developed its own strategy. The first three case-study accounts were discussed and became the basis for the general format for all subsequent reports. The case-study accounts were then synthesized into a single report.

Reflections on methods

The national survey data was very useful to illustrate patterns of utilization of teachers, the availability of science teachers and whether or not they were teaching the subject and at the level for which they have been trained, the situation regarding specialized facilities, their utilization, the availability of equipment, the support available to teachers at school level (from principals, inspectors, heads of science departments, etc.). In

Morocco, it was also possible to obtain information on teachers' practices (in particular, how often they organized practical work or did demonstration themselves). In Malaysia, the survey also illustrated how serious the reduction in science student numbers from rural schools and low-scoring urban schools was becoming and some new empirical relationships appeared that contradicted some common impressions (not least that high-achieving schools tended to have larger class sizes and less extravagant laboratory provision than some low-achieving ones), and representative information was accumulated on several dimensions (for example, the number of teachers who had experienced in-service training) which was not otherwise available.

The value of the survey approach (when it is well done) is that it is relatively cheap to organize and it provides information on the whole country, showing how serious some problems are and what types of school (urban/rural, in which region) are primarily affected. If the sample is representative, it allows generalizations with some certainty. However, some of the issues investigated through the survey proved unsatisfactory. Establishing enrolment patterns and transition rates was complex since significant student transfer can take place between schools. Expenditure data were often uninterpretable since no simple rubric could be devised that was understood in the same way by different accounting officers in the schools.

Contrary to some expectations that the case-study work might prove expensive and add little, it did in fact provide a rich source of insights. Although the problems encountered were indeed specific to the schools studied, there is no reason to believe that these schools were atypical. Thus, what was learnt was at least strongly suggestive of general issues. Many of the things that emerged could not readily have been discovered in other ways. Insights into teaching and learning practice were a powerful indication that learning practices did not always follow curricula intentions, and that the reasons were not usually through lack of resources. Much equipment was infrequently used judging by its condition though it was available. The frequency with which rural school principals were away from schools on legitimate business elsewhere was surprising. Also, the relative neglect of science teaching for non-science students was noticed. The relative paucity of library material in the national language on science at an appropriate level for most secondary school students was noted.

If school case studies are to be used it is critical that appropriate training is provided to field workers. The risk is that the naive case-study worker, especially if lacking the experience of having worked in a school environment, will simply not capture insights beyond the superficial.

Some of the techniques of case study – e.g. checklists – can be used with relatively little preparation providing they have been well conceived in the first place. Others – e.g. interviewing and classroom observation – depend on sensitivity to context, intelligent probing, consistent pursuit of corroboration of views expressed by individuals, classification of data in meaningful ways, and skills of prioritizing important perspectives that emerge. Any attempts to assess qualitative attributes – attitudes, school climate, morale, perceptions of difficulty – depend on experience with relevant social science methods and at least some theoretical understanding of problems of measurement and evaluation of social, psychological and organizational attributes.

Poor interview technique can lead to the pre-existing views of the case-study worker simply being projected back by respondents as a result of leading questions. A good example is where science teachers may be asked to comment on their teaching methods in the full knowledge of the approaches they are supposed to be using. For example, if an interviewer asks:

"Modern science teaching is designed to be student-centred and activity-based and teachers are expected to undertake group work in practical activity in science. You have attended an in-service workshop on how to organize practicals. Would you say your teaching is child-centred and involves group work?"

With this kind of invitation only the most courageous teachers will disagree, whatever their actual practice. More experienced interviewers would probably approach this question by asking specifically about a recent lesson before discussing more general practices:

"Tell me about the last science lesson you taught. What was the purpose? What did you do? What did the students do?" They might subsequently ask for generalizations: "Do other teachers approach lessons the same way? Do you ever use group work? Please give me an example".

Similarly, poor observation technique may lead to a focus on what the teachers do at the expense of what learners experience. It may focus on details irrelevant to the main issues of interest. It may be unsystematic and casual when the need is to gather a picture of common teaching methods and learning activities. It may ignore valuable sources of corroboration – e.g. the record of work in students' books that may provide an indication of the pattern of typical lessons over time and of how much material is assimilated.

Conclusion

Secondary science is a key area of the school curriculum that plays a crucial role in the development process. This makes it critical that flows of information to decision-makers are improved, costs are constrained to those which are justified by effectiveness, and that investment does result in demonstrated learning gains. The best way to achieve these outcomes involves a number of steps.

First, it is desirable to institutionalize periodic review studies of science education which cover the kind of issues raised in the baseline studies undertaken, and collate data on destinations and projected needs for science-qualified school leavers.

Second, cost data should be accumulated which allow monitoring of excessive costs and skewed investment patterns not associated with gains in access or successful participation.

Third, consideration should be given to how best to use existing assessment data to monitor learning achievement, and where these do not exist, to institute some regular method of obtaining indications of learning achievement which can be linked to costs.

The results of employing these planning methods have two audiences – those who take responsibility for the implementation of policy and the policy-makers themselves. The first group, usually consisting of permanent officials and professionals in ministries, need information presented in forms that accurately represent participation in, and the condition of, science education, so that adjustments can be made which reflect policy and represent value for the public money invested.

Insights from the planning process need to inform the generation of policy and to ensure that an appropriate priority is given to science education. Depending on the circumstances this may be achieved through task groups which draw attention to pressing problems where new policy is needed, consultative groups of stakeholders that can contribute to the evolution of policy, or special initiatives to promote science education where participation and quality is lacking. This may be seen to advocate the valuing of this area of educational investment above some others. The IIEP analysis suggests this is often defensible.

Chapter VI
Conclusions and recommendations for planners and decision-makers

1. Introduction

Since the mid-1960s, the world has undergone enormous changes and a phenomenal development in science and technology has taken place. New techniques, new production and consumption patterns have altered, in varying degrees, day-to-day life almost everywhere, even in the most remote rural areas. In this respect, education represents a powerful tool for preparing a future of sustainable economic development. Educating people in science and technology, helping them to understand their environment and act upon it, has become a very powerful instrument to promote a strategy of sustainable economic development, fight against poverty and bring about social welfare. At the same time, the training of high-level and middle-level manpower in science is essential for a country which has to use, maintain, buy or produce technology. Hence, investment in science and technology education is widely regarded as essential and many countries have been consciously pursuing this as part of their educational development policy. IIEP research has analyzed some of the forms this investment takes, has provided insights into the current status of provision at secondary level in a number of countries and has identified key problem areas for careful consideration. In spite of much investment in science education, some of the results of research are not unlike those found in the 1970s. Why is it that many of the promising advances that have been made over the last three decades have not led to more progress in improved teaching and learning and wider scientific literacy? In this concluding chapter, some of the most important findings are highlighted, and some generalizations are offered that it is hoped will focus attention on critical issues for planners.

No general prescription that is universally valid can be offered, since conditions and objectives vary so widely. The policy implications of the various research projects have to be interpreted in the light of specific country circumstances. There are a number of key factors which would

200

seem likely to influence the choice of the strategies for cost-effective science education provision and the conditions under which they could be implemented. These factors include:

- the proportions of students currently enrolled at different educational levels, and the proportions of those who are enrolled who specialize in science;
- the degree of financial constraint on investment in secondary education (i.e. scope for additional investment);
- evidence of shortages and surpluses in supply and demand for science-qualified school leavers at the next level of education and in the labour market;
- the existing level of disparity in male and female enrolments (and/or between other significant groups/regions) in science at different levels;
- evidence on the levels of performance in public examinations and patterns of achievement in science education;
- the historical patterns of provision (existing policy and practice on streaming and selection, curriculum, resource allocation for science);
- the national development policy, which may anticipate more or less scientifically and technologically dependent development investment strategies.

Bearing these variables in mind the rest of this concluding chapter seeks to provide a framework for debate within which policy decisions can be made and appropriate planning and monitoring systems developed. It is based on the main results of the IIEP research project, which included studies on the condition of science education in some 15 countries, state-of-the-art reviews of the literature in Anglophone and Francophone countries, in-depth case studies on the state of science education in two middle-income countries – where the teaching conditions are reasonably satisfactory – and a series of monographs and position papers on specific issues. The main results of our investigations and research work are summarized below, followed by the conclusions drawn and an outline of major policy options.

2. On investment in science education and economic development

The data collated on GNP per capita and the number of students per 10,000 population indicate a loose association between investment in science education and indicators of economic development. The conclusion that can be drawn is that in most cases such investment is a necessary but not a sufficient condition for growth. Although anomalies appear in the data reported in Chapter I – for example, Eastern European countries invested heavily but did not experience growth over the 1970-90 period – these do not undermine the general proposition. The opposing case – i.e. that it is possible to sustain high rates of growth without investment in science and technology human resources and without an adequate stock of related skills in the labour force, is not advanced in any of the literature reviewed, or supported by evidence. Amongst the other conditions that may be necessary to reap the benefits of successful investment strategies in science education, the following deserve listing: a sound macro-economic policy, political stability, a competent and stable public administration, an efficient allocation of resources used in production, a commitment to technologically-based innovation, and appropriate incentives to enterprises to become increasingly competitive.

Evidence suggests that many countries have underestimated the time it takes to acquire a mature and appropriately orientated scientific and technological capability suited to the economic and social realities of production and the delivery of valued services which improve living conditions. Many countries may also have underestimated the need to spread scientific literacy throughout the general population underinvesting in primary and lower secondary education as a result.

The conclusion was reached that the most attractive science education policies, which will contribute most to development, are those that place their stress on two overriding objectives, these are:

- providing basic science education for all (desirably including a technological element) to give a secure foundation to the largest possible group;
- progressively investing in more specialized science education at higher levels (in the upper secondary cycle and in higher education).

These objectives are sometimes placed in opposition to each other as if they were alternatives. In our opinion, this is not justified since neither objective alone satisfies the needs to disseminate scientific and technolog-ical literacy throughout the population and to provide an adequate supply of professional and sub-professional employees in science-based occupations. Many development-related activities benefit from a systematic general understanding of science and technology in society and in the workforce, many also depend on specialized expertise to translate ideas into valued products and services.

The planning questions that arise from this proposed policy orientation are elaborated in later sections. Before turning to these one needs to look at some features of the supply and demand for qualified secondary science students. These students are the most expensive to produce and arguably the most critical to successful investment strategies for development.

3.   On supply and demand

There is a widespread general belief that scientists, engineers and others with technological training are in short supply in most developing countries and that as a result they occupy a favourable position in the competition for job opportunities. What evidence there is suggests that this is indeed often true and that their unemployment rates tend to be lower than for those who specialize in other subjects.

Closer examination of the data suggests that some countries may have begun to produce more science graduates, particularly pure science graduates, than can be absorbed into jobs that reflect particular subjects of specialization. In most developing countries, government is the largest employer of science-qualified personnel. Where public sector austerity has restricted growth in state employment, and pure science graduates outnumber applied science graduates, oversupply is a real risk. The largest demand for pure science-trained graduates usually comes from the teaching profession. Opportunities outside this field may be limited unless graduates are trained to be more versatile, unless they acquire applied competencies and unless they adopt more flexible approaches to the use of their skills. Clearly, there is scope for substitution between pure and applied sciences. It is in the latter that shortages remain fairly endemic, putting aside the specific requirements of the teaching profession. A conclusion is that, especially at higher levels, more students may need to be attracted into applied science programmes, as far as resources permit, and enrolments in pure science limited to those that have some prospect of being absorbed into teaching and research institutions. The case for this is greatest where resources are most scarce and where pure scientists have no special comparative advantage in the labour market.

It was virtually impossible to evaluate information on the employment and work opportunities for science-trained secondary school leavers. What little there is at a country level is piecemeal, and often unreliable. Though there is evidence of growing unemployment of lower secondary school leavers, the blame can hardly be put on education and on science education in particular. At upper secondary level, existing data did not establish whether science-qualified school leavers are better or worse off than their peers. Even if it could be shown that they are more attractive to employers, it would be difficult to say whether this is because of the knowledge and skills they have acquired or because they may be from a selected group of students with higher than average ability. The partial data from Malaysia, where one of IIEP's case studies was conducted, indicate that some employers preferred science school leavers, even for training in areas like accountancy which is not scientifically based. However, there were reservations about the quality of recent school leavers' capabilities, suggesting that even if the demand was strong, this did not indicate satisfaction with the skills and intellectual abilities associated with science school leavers.

In most of the Francophone countries studied virtually all science-stream graduates continue into higher education, mostly on science and technology courses. In some countries, output from secondary school is greater than the places available in science and technology in higher education (e.g. Burkina Faso). Such school graduates are absorbed into subjects that do not have a science base. The same is happening in Morocco but for another reason, i.e. the lack of employment prospects for pure science graduates. If it can be shown that their subsequent performance is inferior to that of entrants from non-science streams – as is apparently the case for undergraduates in economics – adjustments may be called for in the selection policy into the science stream and the numbers admitted. It might also be appropriate to review the policy options regarding the curriculum in order to reduce the degree of specialization and/or make science programmes less theoretical and more applied.

In some Latin-American countries and in Malaysia, school output of science specialists is comparable with, or less than, the number of places available at higher levels. Where this is true, higher and further education institutions have to recruit students who are not well qualified in science (Malaysia), reduce enrolments (Chile), or organize specialized recovery courses at university level (Argentina, Botswana and, to a certain extent, Mexico). Postponing specialization in science and technology until post-secondary level may be attractive if good-quality basic science education can be widely provided. If not, some amount of earlier specialization, sufficient to secure the supply to higher levels, seems preferable to high-cost recovery courses which teach material that could already have been mastered more easily in the school system at lower cost.

This supposes that secondary school pupils can be encouraged to enrol in science streams or options. This is not always the case – Malaysia and Chile are examples of countries where fewer and fewer students want to enrol in science in spite of the existence of favourable employment opportunities. Market signals for science-qualified students can be ambiguous. Whilst the salaries of applied scientists and engineers are generally well above average, salaries of pure scientists are often lower than those for professionals in other fields (business, accountancy, economics, law). A major explanation is that a usually much greater proportion of pure scientists are employed by government (especially as teachers). Even when teachers are removed from the equation, the lower salaries of pure scientists will reflect the lower salaries of professionals in the public sector rather than those who are self-employed consultants or working for large (and frequently multinational) companies. Similar patterns may be observed at the technician level.

Analyzing the demand for science education, and identifying the appropriate policy conclusions that can be reached, are therefore problematic. Several different situations can exist, as shown below in *Table 6.1*, and each may have different causes and suggest different policy options.

In each case, demand has to be coupled with analysis of the supply of opportunities. Thus, many students may want to study secondary science (strong student demand), but few places are available (weak supply). It does not follow that strong individual demand should always be reflected in increased supply of places (unless the most extreme forms of neo-liberal reasoning are employed). Independent judgements have to be made of both what can be afforded in terms of the supply of places and what it would be in the national interest to provide. Market signals are important to analyze but they only provide information on a short-term basis. Decisions on supply, especially at higher levels, have to take a medium- to long-term perspective.

The extreme cases in *Table 6.1* are fairly clear-cut. Consistent strong demand suggests under-supply; consistent weak demand suggests over-supply, assuming labour market signals are consistent with this explanation. The intermediate cases are more difficult to interpret. They may suggest over- or under-supply at secondary or higher levels; they may reflect an over-emphasis on science in the curriculum (or required combination of subjects), although specialization is not required for subsequent study or work; mismatch may also arise because perceptions of opportunity are falsely constructed in the absence of reliable information.

Table 6.1   Demand for science education

Demand for science places at secondary level	Demand for science/ applied science places in higher education	Demand for science– specialized workers and professionals in labour market
Strong (e.g. Thailand)	Strong	Strong
Strong (e.g. Morocco)	Strong	Weak
Strong	Weak	Weak
Weak (e.g. Malaysia)	Strong	Strong
Weak	Weak	Strong
Weak	Weak	Weak
Weak	Strong	Weak
Strong	Weak	Strong

Increased output seems most difficult to justify where labour market demand appears weak, except in those areas where growth is expected to be rapid and anticipatory intervention is consciously intended (as in the case of education and training to supply a rapidly growing economic sector). Where demand is strong at the higher education level but weak at the secondary level (as in Malaysia and Chile), there is a risk that the supply of qualified applicants may fall below the numbers needed to fill available places. Steps may need to be taken to give a more positive image of scientific activity and incentives need to be offered to encourage more students to choose science (especially applied science/technological) options. In other cases, incentives may have to be offered to the responsible authorities so that they provide more science options.

In most of the countries studied, comprehensive and reliable information on opportunities for science-qualified school leavers for higher studies and in the labour market is not available. Choices by students and parents are more likely to be made on the basis of perceived difficulty, personal preference, hearsay of labour market experiences, and a narrow range of stereotyped occupational futures, than on the basis of realistic appraisal of options based on sound information. Patterns of effective demand are likely to be influenced by a realistic appraisal of both opportunities and the chances of taking them up. Efforts should be made to make more of this kind of information widely available.

Before ending this section, it may be useful to make a few remarks on the specific case of low-income countries, African ones in particular. Indeed, many African countries have a very low proportion of their population enrolled in science and engineering at tertiary level. Costs are typically very high and can absorb very disproportionate amounts of the education budget. The results of past investment have been to a certain extent disappointing for a variety of economic, political, and institutional reasons. Economic migration of the qualified has often neutralized effects on the stock of nationals employed; poor and unstable working conditions have encouraged talented scientists and engineers to seek other forms of employment; salary structures linked to those of the mass of public servants, who may be in specialities where there is no shortage of supply, have penalized those with rarer talents.

If these problems were easy to solve they would not recur. Some scope is seen, under different circumstances, for one or more of the following to improve output and retention:

- active localization policies that require counterpart training and penalize persistent use of expatriates, coupled with performance safeguards;
- comparability pay exercises for scientific and technological professionals in the public sector with those in the private sector;
- improved information about career opportunities and access to higher studies in science-related areas;
- strategic use of training in metropolitan countries and other developing countries where there are clear cost and quality gains;
- direct incentives to study science;
- action to increase female participation at all levels of education and raise their interest and achievements in science (see below);
- interventions designed to improve science achievement (better materials design, enrichment for difficult topics, focus on under-achieving students).

4. On participation

Providing science education at lower and upper secondary level has been argued to be of crucial importance to the development of human resources for several reasons:

- it consolidates and strengthens what pupils have learnt at basic education level;
- it provides a sound basis for the science education of those who will become teachers of the next generation of primary school students, some of whom will study up to higher levels and become specialists who teach at secondary level;
- it offers a preparation for middle-level scientifically and technically based workers, many of whom will take technical and vocational courses subsequently, who will benefit from a secure foundation in basic concepts;
- it facilitates the selection and training of the future scientific and technical élite who will proceed to university level and professional jobs.

The implication is that all students should continue to study some core science throughout the secondary cycle. The reasons for advocating compulsory core science, in addition to those mentioned above, are that:

- science should be regarded as an integral part of general education, since it represents distinctive ways of knowing and interpreting the natural and constructed environment;
- compulsory core science will enhance the participation of those who might otherwise choose to opt out, with a consequent narrowing of their career opportunities and loss of national talent (girls, rural students, etc.);
- those who develop an interest and ability in science during secondary school and are late developers will not be excluded from science subsequently through premature streaming and tracking.

It was found that provision and participation in some form of science education at lower secondary level was almost universal amongst those enrolled. It is to be stressed however that enrolment rates in lower secondary education differ widely among the countries covered by the IIEP research project. Participation at upper secondary level is even more disparate and many different patterns exist, including those where no science is studied by some (i.e. students have the possibility to opt out of science completely).

5. On gender

The review and research data lead to the conclusion that observed differences in participation in science education by gender are primarily a function of cultural preferences and option choice systems. Whatever differences may exist in cognitive style and capability are too small to create these disparities. The fact that differences are not stable between countries (some have no differences at secondary level, others have large ones) and that differences change over time (girls now outperform boys in England at the school-leaving age) support this view.

A general conclusion from the studies is that relatively low participation in science education by girls at secondary level in some countries is both inefficient and inequitable. It implies talented girls are lost to science-based careers (particularly undesirable where shortages are chronic) and that more could participate if given appropriate encouragement and support. The raising of girls' achievements in science at lower and upper secondary levels (non-science stream) is of equal importance for development and the fight against poverty. It would also help to reduce fertility rates where these are excessive. Some of the following steps could be taken to restore the balance between boys' and girls' participation:

- ensuring that as many places are available for girls to study science as boys;
- promoting role models of successful female scientists;
- increasing the proportion of female science teachers;
- creating girl-only science classes;
- constructing gender-neutral tests;
- developing gender-neutral learning materials.

6. Specialization

Specialization in science can be defined in a wide variety of ways. A rule of thumb is to apply it to curricula options where more than 25 per cent of curriculum time is spent on science subjects (including technology or applied-science courses). There is no simple way of determining the optimum numbers of students who should be encouraged to specialize in science at the levels where some specialization is permitted. The decision will depend on the contextual factors identified above in the introduction, which vary from country to country.

In most developing countries it would seem that a minimum of 20 per cent of upper secondary enrolment should specialize in science unless there are very strong indicators of over supply or where a substantial science curriculum for all exists. This minimum may be considered too low if a high number of those who specialize do not qualify, and if a significant proportion opt for other specializations at higher levels. If technically based studies are included in this calculation, the minimum proportion should be increased to 30 per cent.

Conversely, where more than 50 per cent of upper secondary enrolment is specialized in science the proportion may be higher than is justified by the associated costs. It is unlikely that such high levels of output will be absorbed into related further studies and labour market opportunities in a sustainable way. Unless general advantages are seen in the rigours of successfully undertaking intermediate-level science courses, it may be that scaling down of enrolment should be considered, especially if this is predominantly in pure science options.

Some countries (e.g. Malaysia, Korea, Nigeria) have created special institutions to concentrate resources on the highest quality of science education that can be provided to generate a core scientific and technological élite. In Francophone countries, special tracks exist where the best students in mathematics and physics are trained, but which are not necessarily available in all schools. Graduates of these institutions are supposed to become the future élite of the country. The proportion selected for these very specialized educational experiences are likely to be small. High unit costs, coupled with limited opportunities for employment at the professional level, suggest limits of around 5 per cent as being plausible. Much less than this suggests the impact on professional employment may be marginal; much more suggests dilution of the special effort in favour of preparing students for a much broader range of scientifically based occupations. If special science-based schools are favoured as a policy option, it is essential that these are coupled with the monitoring of the outcomes to ascertain whether much higher costs are justified by levels of achievement and by patterns of absorption into higher studies and jobs. The impact of such special arrangements on the rest of the system should also be monitored (identifying possible perverse or negative effects).

Many developing countries continue to experience more acute shortages of scientific and technically competent staff at the middle levels of the labour market than at the professional level. At some point, selection has to occur so that students can be tracked into appropriate education and training opportunities. Where those leaving technical and vocational institutions at the upper secondary level can be shown to have a greater attraction to potential employers than those from the general school system, there may be a case for expansion. Where this is not the case, the best preparation for the labour market may remain in mainstream schooling which can provide a secure basic science education with some technological orientation. In low-income countries and in those countries where participation at secondary level is still limited, it is attractive to see technological concerns treated as an integral part of science education rather than as a separate subject. Many learning objectives of technology overlap those of science education. Integrating technological concern into the science curriculum will almost certainly be cheaper than offering technology as a separate subject. If science subject matter is to be useful beyond school, it should have an applied and technological flavour. It would seem difficult to argue that technology should be studied without a grounding in basic science; the converse is rarely argued.

7. On science curricula

This research was not primarily concerned with curriculum issues. It is inevitable, however, that these have to be addressed, since they carry with them implications for costs, internal efficiency and quality, which are the legitimate concern of planners and policy-makers.

In some countries, science constitutes a single subject in lower secondary curricula; in others, two or more science subjects are taught separately, either concurrently or in sequential order. The most common patterns in countries with an Anglophone tradition now incorporate single 'integrated' curricula that draw on the three major science disciplines and sometimes include aspects of agricultural science, earth science, health education and environmental studies. Francophone countries often integrate two subjects and teach physical science and life science. To others may be left the resolution of the methodological problems of deciding whether integrated approaches are superior according to some criteria rather than the alternatives. It is noted that the IEA data show that high levels of science achievement can be attained through a variety of patterns. The concern is that whichever option is chosen the most appropriate costs are factored into the decision. Single integrated courses should be cheaper to deliver at equivalent quality, since they offer opportunities to simplify textbook production, teacher training, staffing, timetabling, laboratory provision, and examining, and enjoy economies of scale. By implication it is usually preferable to incorporate material from minor subjects (agriculture, health, environment, earth science) into one (integrated) or two (life science, physical science) courses rather than to attempt to teach them separately.

Similar arguments can be applied at upper secondary level, where it is much more common to teach science in some combination of general science, especially to art students, and specialized separate subjects to science students. The more options there are, the more the costs and logistical complications of provision will escalate. Everything being equal, costs (and problems of teacher deployment) will be reduced if only one to two science disciplines are offered. It is noted that learning objectives across the three main sciences (physics, chemistry and biology) and the minor related subjects have more in common than otherwise at all levels in the secondary school. The planners' questions must be to ask:

- What is the special value of particular options that justifies the additional costs?
- What are the science learning objectives that distinguish these science subjects so clearly that separate teaching is needed?
- Can unique learning objectives be incorporated into a broader based and more applied secondary science curricula?

The IIEP studies indicate that science curricula are often overloaded with content (teachers feel unable to complete the syllabus) and that factual knowledge is frequently valued to the exclusion of higher-order intellectual skills. Recent research in the United Kingdom on cognition suggests that cognitive gains in science amongst secondary students can be accelerated by appropriate curriculum interventions designed to promote the thinking skills and analytic approaches characteristic of science. If these results can be replicated in other country contexts it is strongly suggestive of the value of shifting the emphasis in curriculum development away from extensive content and superficially understood experimentation towards the intellectual processes that are generally considered to be of most enduring benefit. In brief, if simple experiments can be designed to generate 'cognitive conflict' and systematically lead students through a reasoning process, they may prove very effective at promoting reasoning abilities and analytic interpretation. More use should be made of these approaches in curriculum design. In this view of the curriculum, content remains important (since no process skill is really content free) but it is subordinate to mental engagement. Overloaded curricula could be culled and practical activity should be focused on the development of the mental attributes it is supposed to develop. Such an approach would also reduce the number of costly experiments to be included in the syllabuses.

Some parts of the work supported the view that curriculum presentation adversely affects particular groups (e.g. girls, language groups, rural students). There were differences in the material boys and girls considered difficult; in some cases, assessments tested language competence more than scientific ability; rural students can actually outperform urban students on items that are presented in contexts more familiar to them; textbook images still over-represented some stereotypes and undervalued available role models. Where this is so, it should be the subject of curriculum development work.

Observations of science curricula led to the conclusion that it was rare for different pathways through materials for students with different needs to be made explicitly available. Where educational infrastructure is poorly developed this may be an unrealistic aspiration. In better resourced environments, quality may be enhanced by recognizing that students learn at different rates and have different needs. Core courses do not preclude optional units which contain enrichment and remedial material and which can reflect regional variations in relevant science (dry/wet zone, agricultural/coastal, etc.).

Mathematics has a close relationship to science. Performance in mathematics is often correlated highly with science performance. When curriculum options are considered it is essential that mathematics is included as a compulsory subject. The possibility of opting out of mathematics that exists for secondary students in some countries (e.g. in Thailand at upper secondary level) needs to be seriously reconsidered. So also do links between the mathematics curriculum and science options so that common conventions are used and the sequencing of material is appropriate.

Finally, it was noted that assessment data rarely seemed to be used as inputs to the curriculum development process. Despite the insights they can give into areas of difficulty and parts of curricula that disadvantage particular groups, and despite the importance of examination performance as a determinant of curricula in action, analyses of assessment data for curriculum purposes seemed rare and often findings concerning performance were restricted to those within examining agencies rather than shared with curriculum developers.

In summary, it is suggested that:

- the benefits from integrating science into a single subject at lower secondary are considerable and the disadvantages marginal;
- science options at upper secondary should not be allowed to proliferate unless convincing cases based on unique and valued learning outcomes can be made;
- content should be reduced wherever possible in favour of more emphasis on systematic attempts to develop higher-order cognitive capabilities than recall;

- content should be selected taking due account of material most likely to be available to the majority of students studying particular programmes;
- presentation of curriculum materials should be free of unnecessary stereotypes and should draw examples from the full range of backgrounds from which students originate;
- flexible curriculum design should be employed to allow local decisions by teachers of optional content to reflect local circumstances in the context of common core material;
- the mathematics curriculum should be closely related to the science curriculum;
- more use should be made of assessment data to inform curriculum development targeted on reducing learning difficulties and improving the performance of under-achieving groups.

8. On learning and teaching

Learning and teaching conditions vary very much from one country to another. They also vary considerably within countries. In the study, African countries tend to have the most problems providing satisfactory teaching conditions in the majority of schools. Typically:

- many schools have high class sizes, in some cases as many as a hundred pupils, with few possibilities of meaningful group or individual work and little direct contact with teachers;
- some countries still have substantial numbers of expatriate teachers who may not speak local languages, may have transient knowledge of the curriculum, use locally unfamiliar teaching styles and create problems related to high rates of turnover;
- many schools in rural areas experience high rates of science teacher turnover and consequent lack of continuity. Shortages of science teachers are also common to the extent that untrained teachers are employed to teach science or that science is not taught at all in certain schools or to certain classes.

All countries however experience in one way or another unsatisfactory teaching conditions:

- Teacher deployment is widely problematic, even in middle-income developing countries where shortages in some schools appear along with surpluses in others. Many science teachers do not teach the subjects for which they were trained whilst science is taught by others not trained in science.

- Relatively low salaries, whose real value has been eroded over time, are often cited as a cause of lack of teacher motivation. The situation is particularly critical in some Latin-American countries, where teachers end up teaching an excessive number of hours to many classes to make up for their loss.
- Teaching methods are widely criticized for emphasizing recall and rote learning rather than a journey of enlightenment and a reflective experience which generates useful knowledge and skills.
- Laboratories for science are widely underutilized, or wrongly utilized, i.e. they are used for traditional whole class teaching. Maintenance of equipment and refurbishing of consumables is a major problem.
- Lack of equipment and examination pressure are often cited as constraints on teaching and learning.
- Striking inequalities between schools are a major characteristic of many countries' systems. Lack of standardization, decentralization and privatization have contributed, in some countries, to the lowering of standards and increasing disparities between the best and the poorest schools.
- Within schools allocative decisions can result in perverse patterns of investment.

All of these problems have their specific cause and possible solution in each country. Comments are made on just three.

First, science teacher deployment is a critical issue. It represents by far the largest recurrent cost of teaching science. Yet, often teacher posting is not driven clearly by need. District- and Provincial-level planning data may not be able to identify the science teacher establishment in a particular school and are even less likely to be able to relate this to the number of students actually studying science. Formula-based staffing that relates science teacher posting to indicators of numbers of science students (or, at the very least, examination entrants) is unusual. Without it existing inefficiencies will persist.

Second, although lack of material resources (equipment, consumables, laboratory environment) can be a very real constraint on how much science can be taught, there is evidence that these reasons are given even by teachers in relatively well-resourced schools. Teacher motivation, skill in planning flexible and creative lessons, and lack of understanding of curriculum objectives are all likely to be contributory factors in determining why so much of the science that is taught appears to diverge from the expectations of curriculum developers. On occasions, it may also be these expectations which are at fault. Curricula designed for high-cost urban schools with students from homes rich in educational resources may indeed be inoperable in typical schools in poor communities.

Third, although what students actually experience in science education is largely determined by school-level decisions about who teaches them under what conditions and with what resources, not much is known about allocative practices within the school, or between types of schools. It is noted that often grossly disproportionate allocations are made for the study of science by small groups of senior students at the expense of lower grades. Lower class sizes are often excessively large, and taught in the poorest conditions by the least-qualified teachers. Timetabling of science teachers can result in very low student-period loads. High-level examination results are often pursued at the expense of investment in the teaching of science to younger students. There seem to be few institutional mechanisms which discourage these practices.

9. On learning materials

Available curriculum materials vary in quality and relevance from the excellent to the obviously inadequate. Availability also varies from adequate to virtually unavailable. In some countries, 'unofficial' curriculum material – workbooks, examination guides – are more popular with students (and sometimes teachers) than official curriculum materials. The range of materials can be narrow – a single textbook – to comprehensive – students books, worksheets, teachers' guides, enrichment material, adequate library stocks. In some countries, a nominal charge is made, and in some, the full commercial cost is charged. Free provision is the exception in developing countries in secondary schools.

Although it is difficult to draw general conclusions from the research, *Table 6.2* suggests a starting point for country-specific analysis:

Table 6.2.   Policy options concerning curriculum materials

Quality and relevance of existing materials	
High	Concentrate on supporting effective use of materials through in-service and school-orientated support
Low	Invest in curriculum development
Availability of basic texts	
High	Consider increasing the range of materials and the possibility of selective recovery of some of the costs
Low	Reduce costs; invest in effective distribution systems; subsidize purchase and delivery to underserved schools
Use of alternative curriculum materials	
High	Analyze why 'official' materials are not preferred; invest in curriculum development and possibly improved assessment systems if examination-orientated learning is a major reason. Consider revised system of textbook production that is more sensitive to effective demand for text materials
Low	Explore the options to extend the range of materials available
Range of printed materials	
Narrow	Invest in curriculum development of enrichment materials, teachers' guides, and other learning aids
Wide	Provide advice on coherent choice of core materials
Cost	
High	Reduce cost to levels that are affordable by providing selective subsidies and encouraging regional co-operation.
Low	Consider selective cost recovery.

10.   On practical activity

There is extensive evidence from the case studies and those that have been reviewed by IIEP that much practical activity provided in science education falls short of practices that would justify the additional costs of universalizing high levels of physical provision in resource-poor countries.

There are several reasons for this:

- much practical activity overvalues actions at the expense of thinking skills;
- there is little evidence that practical work, as it is commonly undertaken, has a positive effect on measured achievement;
- expensive laboratory environments may frequently be used for general class teaching;
- the disposition of laboratory facilities is often very uneven;
- utilization rates for expensive equipment may be low, effective maintenance arrangements are rare, and costs for consumable material may be unsustainable.

The picture that emerges is one in which practical activity and laboratory-based science is conducted in an intellectually challenging way in a minority of cases. More often, practical work is presented in ritualized ways which follow stereotyped steps. Students have little opportunity to identify their own problems, play a role in the development of appropriate experiments, and collect and interpret data themselves. Their learning may be 'minds off' rather than 'minds on' in the sense that the cognitive skills associated with problem solving take second place to the following of instructions, which are often poorly understood.

The tentative conclusions that can be drawn, which are of greatest relevance where resources are scarce, are that:

- Wherever possible, practical work should be designed with costs in mind to make appropriate experiments available to relatively poorly resourced schools which have large classes.
- The quantity of laboratory-based work should be considered in the light of the learning gains associated with it. It may be that fewer extended but simple experiments, along with simple practical activities which may not be experimental, are a preferable and more cost-effective option than curricula that assume experiments take place virtually every period.
- Expensive and rarely used equipment should be eliminated from the science curriculum wherever possible.
- Alternatives to laboratory-based work – thought experiments, demonstrations, simulations, video presentations, excursions in the natural environment – should be considered as an option to individual or group class work.

11. On equipment

The cost of equipment depends on what is specified and where it is produced. National lists of the minimum necessary equipment to teach science should be constructed where these do not already exist, and should be restricted to that which is essential.

In summary, the findings suggest that:

- high-cost individual items should be avoided, especially if infrequently used;
- imported equipment should always be assessed to determine whether local alternatives of adequate quality can be produced;
- appropriately designed science kits should be considered as an alternative or as a supplement to the existing equipment base where cost precludes comprehensively equipping all schools;
- if kits are deployed, advice on their use and arrangements for replenishment should be part of an implementation package.

12. On laboratories

Costs of laboratory provision vary across an enormous range relative to the cost of ordinary classrooms and may reach ratios of 5:1 or more. Where costs are many multiples of normal classroom costs and provision is well short of the number needed for all secondary schools, lower cost alternatives should be considered for the short to medium term. The most obvious option is to agree on the design of science rooms, i.e. ordinary classrooms that have a selected range of basic facilities adequate to teach non-specialized science through to the end of upper secondary school. This should be possible at no more than double the cost of ordinary classrooms.

It is suggested that:

- Laboratory costs should be a relatively small multiple of classroom-building costs in all but the most well-resourced school systems.
- The design of laboratories should, wherever possible, seek to provide a basic range of services that can be sustained and which are appropriate to location, with adequate lighting and ventilation.
- Secure storage should be incorporated into the design, and consideration given to safety.

- Where space permits, furnishing should allow individual and small-group work.
- Provision should allow for visible demonstration work.
- Science rooms, and multipurpose specialized rooms with science kits, are an acceptable alternative to laboratories at lower secondary level. Even at upper secondary level, most learning objectives can be achieved through work which does not require laboratories. Multipurpose laboratories are a reasonable solution in resource-poor systems.

13. On language

For historical, cultural, political and other reasons, some countries included in the project teach science in a foreign language, while others use one, or more than one, of the national languages. In the first type of country, the learning of science (and other subjects) in secondary schools seems to raise more difficulties than in those where pupils use their mother tongue (or a widely used national language). More emphasis on adequate language fluency in the selection and training of teachers may help overcome some of these difficulties, but could also exacerbate teacher shortages. Dual-language textbooks and teacher guides do exist and could be helpful in ameliorating problems of science teaching in a foreign language, but require additional investment.

In most countries which use a national language not spoken widely outside the country, science instruction at secondary level has different problems which centre on the amount of written material available in addition to the textbook, adequate technical quality in translation, and the problem of transition to higher studies. Although commonly national language textbooks are available, often little else with a science orientation exists. The translation of scientific terminology has to be standardized at a national level and a full range of terms recognizable to mother-tongue speakers created. This may be a major undertaking. With a few exceptions it seems obvious that at tertiary level, where science becomes more specialized, access to a major international language is essential for science and technology students, whether or not they are taught in it as the medium of instruction. It is beyond the means of all but the richest developing countries to translate a sufficient range of material to provide a good grounding in science at undergraduate level. Lack of fluency in an international language will preclude study at higher levels. At secondary level, especially lower secondary level, on the other hand, the use of the mother tongue or of a widely spoken national language has major advantages.

It should be remembered that language problems in the teaching of science have several dimensions. Some stem from a general lack of understanding of the medium of instruction, whatever the subject matter. Others relate to misinterpretation of specific terminology. Finally, the language of scientific reasoning (cause and effect, hypothesizing, inference, probability, etc.) may have no direct equivalents in some languages. It is important to decide which of these problems may need attention.

It is noted that:

- language policy in the school system will be determined by considerations much broader than considerations of the effect on the science curriculum;
- lower-level science is almost certainly best taught in a mother tongue/national language where this is possible and consistent with national policy;
- a decision has to be made concerning the grade level at which an international language may be used and adequate transitionary arrangements made;
- teaching materials should be designed with language issues in mind.

14. On assessment and achievement

In most countries, policy-makers and planners know little about the actual levels and patterns of science achievement at secondary level. International assessment tests record wide variations among and within countries on the levels of science achievement at the end of lower secondary and amongst those specializing in science at higher levels. It is clearly possible for all countries to accomplish high levels of achievement in science amongst small selected groups of students given special treatment. Interestingly, the IEA data suggest that the achievement of the best students is not adversely affected by the increased proportions studying science. That is, the best students do not seem to suffer if greater proportions are enrolled in studying science. Everywhere, the levels of achievement of average and below-average students are cause for concern. The 'yield' of science education for these students suggests that many fall short of achieving modest levels of scientific literacy.

Wherever possible, levels of participation should be considered alongside assessment data that indicate how much is being learnt. Where achievement levels are low, extending participation may simply exacerbate teaching and learning problems and have little impact on the supply of qualified science-based school leavers or on scientific literacy. It is not uncommon to find raw score performance data indicating that 30 to 40 per cent of students following secondary science courses achieve littlemore than chance-guessing levels on objective multiple-choice tests. Where this is true it may be that assessments test schools and teachers rather more than they do students. It is therefore a priority to explore the causes of under-performance in depth before replicating ineffective investment strategies.

IIEP's research re-emphasizes the importance of using available assessment data to explore levels of achievement, areas of learning difficulty, and reasons for relative under-performance in science of particular groups. This kind of data is essential to curriculum development, informed by analysis of performance, judged against learning objectives. Too rarely, does this seem either a starting point or an integral part of the curriculum development process, as noted above.

Examination-orientated learning appears characteristic of science education in many of the countries studied. It is most obvious where there are established national selection examinations which determine life chances for students. The effect these examinations have on the curriculum can be considerable and is well documented. There are options to reduce the adverse effects, which include higher quality item writing, more diverse selection methods (including school-based assessment, where this is viable), movement towards certification rather than selection examinations at the end of a secondary cycle, and a reduction in the number of selection examinations. These have to be balanced against the possible negative effects on motivation, curriculum regulation, and standards of achievement, which may deteriorate in the absence of external examinations.

A particular issue is raised by the continuing attachment of some national systems to metropolitan examining boards and conventions. International arrangements of this kind often involve substantial costs. If these arrangements impose conditions no longer appropriate to national needs and national curriculum priorities, they should be reassessed. It may be that the advantages that can be provided (of security, technically competent examination construction, recognized standards) are worth preserving in some form. However, the potential disadvantages (of cost, misunderstanding of reasonable expectations of teaching conditions for the majority of students, inappropriate values) of continuing such arrangements need careful consideration.

Whatever system of examination is in place, issues of standardization and standards have to be carefully considered. In some countries, science is no more or less difficult to pass than other subjects since results are standardized to ensure that this is the case. In others, science is much more difficult to pass than some other subjects. Often, it is not clear why this should be, since it assumes that science is intrinsically more difficult at a given level than, say, history. In the absence of compelling reasons to the contrary, pass rates in all major subjects should converge around similar averages on national selection tests. It is a separate question to decide what levels of achievement are actually represented by pass (and are indicated by raw scores). Setting standards too high so that most fail to reach them may be counterproductive for classroom morale and meaningful learning.

In summary:

- Assessment data should be analyzed to establish actual levels of achievement and areas of special difficulty.
- The achievement of low-scoring students should be a focus of special interest.
- Opportunities should be taken to make positive use of examination 'backwash' to encourage desired patterns of teaching and learning and learning outcomes above the level of recall.
- School-based assessment is desirable, but may not be feasible, unless fairly high levels of professionalism have been reached amongst teachers and adequate moderating procedures can be employed. Mixed systems which couple external examinations with internal assessments can offer the advantages of both – objective assessment, assessment related to actual classroom experience, the testing of a wide range of science learning outcomes – but may have some additional costs.
- Where relationships with external examining authorities are retained, steps should be taken to develop indigenous capacity to localize examining.
- Science pass rates should become similar to those for other major subjects through the use of standardization.
- Standards should be set which are achievable by the majority of students in average schools; as performance improves, standards should be raised.

15. On monitoring

System-level monitoring of science education performance is widely lacking. The data suggest that many countries do not know how many students are studying how much science for how long. The multiplication of options, and the multiplication of types of schools, may make it difficult, but by no means impossible, to identify how many students study science and to what depth. Planners and managers of secondary and higher education need to have this information at hand.

Without more effort in this direction it will remain difficult to assess the impact of past policy, identify interventions that are more and less cost-effective, and judge whether levels of participation can and should be increased. The research leads us to advocate a range of monitoring practices that could prove beneficial. Where public examination data are available for reanalysis, this can provide a starting point. Where they are not, consideration should be given to monitoring assessments, covering at least a national sample of schools, which can be used to trace the effects of policy, identify changing standards, and locate areas of under-provision.

Monitoring systems can use public examination data to expose elements of system performance (e.g. distributional and utilizational indicators of science education resources, unit costs in different institutions); identify schools where provision is most (un)satisfactory (e.g. through highlighting anomalous performance with time-series data, comparisons of relatively high- and low-scoring schools with apparently similar initial conditions); and assess improvements in levels of achievement over time (analysis of raw score data, measures of achievement linked to participation).

Monitoring systems that are specially designed can capture other types of data on system performance. They may be able to link achievement with school entry and exit data. They can potentially explore a broader range of outcomes (especially those associated with practical work) than can be analyzed through public examinations. They offer opportunities to collect detailed longitudinal data, which are of unique value, but almost entirely absent from the systems studied.

The following points are recommended:

- the analysis of existing assessment data for monitoring. This is very desirable and this needs to be undertaken at a detailed level by technically competent researchers. Crude pass rates are often insufficient to monitor effectively;
- mechanisms to feed information from monitoring research into operational activity. Science support staff should have access to performance data indicating areas of difficulty, etc.;
- where national assessment data are not easily available for analysis, monitoring assessments, conducted on a sampling basis, should be considered;

- longitudinal as well as cross-sectional monitoring. This should include follow-up studies into higher and further education and into employment.

16. On management and teacher support

The management of staff and resources to deliver science education is widely problematic if our research results are illustrative. Usually, those responsible for science on a day-to-day basis in schools will have had little or no formal training in managing a science team of teachers, organizing the work of ancillary staff, and making purchasing decisions with a view to minimizing costs and maximizing learning benefits. They may also have little authority over decisions that directly affect the teaching of science. In many cases, school principals are unlikely to have come from a science background themselves and may therefore lack confidence and competence in decision making concerning major items of expenditure and science curriculum issues. Where adequate training and support is lacking, arrangements should be considered to provide it. Appropriate authority should be invested in senior teachers charged with the responsibility to lead science-teaching teams.

School monitoring and inspection staff should direct special attention to core areas of the curriculum, of which it is argued science should be one. This implies a cadre of staff with relevant skills and experience available, in sufficient numbers to act effectively. The relationships between them and the executing agency responsible for physical provision should be carefully considered if recommendations are to be translated into actions. Wherever possible, the monitoring and support functions of advisory staff should be deployed separately from those concerned with administrative regulation to reduce conflicts of interest.

In some cases, improvement in science teachers' subject knowledge still constitutes a priority problem which needs to be addressed through pre-service training of better quality; in many other countries – where this problem has been largely solved – emphasis needs to be given to the enhancement of subject matter-related didactic and classroom management skills. The training of science teachers in the area of maintenance and repair of simple equipment also constitutes an important area. The latter may be best approached by increasing the share of training on-the-job and in developing more adequate support systems within the school.

We note that the studies have uncovered a paucity of data on the effectiveness of common methods of in-service training and teacher upgrading for science teachers. Although often these are expensive in terms of direct costs and lost teacher time, medium- and long-term effects are usually unknown. Where changes in recommended teaching methods have not taken root despite several cycles of in-service support, there is a pressing need to re-examine the assumptions that lie behind training methods to see if they can be validated. What evidence there is suggests that more extended school or locally based support, and the introduction of new forms of nomination or 'licensing' of teachers, may be an attractive option both in terms of costs and quality improvement. Few written materials are usually available to support under-trained teachers and this would seem a fertile avenue to pursue.

In brief:

- Steps should be taken to improve the quality of management of science departments in secondary schools through more systematic training of senior science staff.
- Consideration should be given to the range of responsibilities to be delegated to science heads of department.
- Principals who are not science specialists should be introduced to key issues in managing effective science teaching.
- Where feasible, advisory staff should be appointed with responsibilities to improve science teaching and learning across groups of schools. Clustering of science departments collaborating on professional development may be one possibility.
- Science inspectors should try and come to an agreement with science staff on targets for improvement and the additional resources needed.
- Where the science knowledge and skills amongst teachers are deficient, up-grading directed towards ameliorating this should be considered; where knowledge and skill in science are adequate, more attention should be placed on assisting with the development of effective teaching methods.
- The identification of cost-effective methods of in-service support within different systems is a priority. More attention should be given to methods that are school based and provide a continuity of support.
- In many countries, there is a pressing need for written materials for teachers to provide a source of advice on the presentation and organization of effective science teaching.

17. Postscript

This book has described a wide variety of experiences on the provision of science education at school level and has highlighted the main issues that will preoccupy planners into the twenty-first century. It documents how widely investment in science education is regarded as a central feature of education and development policy, how richly varied are the strategies employed to improve access and quality, and how difficult it is to meet aspirations to achieve science for all and ensure an adequate supply of science-qualified school leavers to feed higher education and the labour market in many developing countries.

Many challenges remain and some are new. The research has uncovered many promising initiatives. All too often these have the characteristics of piecemeal innovations which have an impact on parts of the human resource development problem in science, but do not form a component of a coherent, sustainable plan for the medium term. Crisis management of science education issues seems more common than planned change over a period long enough to show consolidated gains.

Many of the interventions identified are not new in the sense that they have never been tried anywhere before. This is not surprising given the investment of time and resources in science education over the last decades. There are many examples of the introduction of science kits, low-cost equipment production, curriculum reform programmes and in-service teacher-training initiatives. Our research pinpointed a combination of reasons why such initiatives have not led to more progress in improved teaching and learning and wider scientific literacy, amongst which are:

- A lack of resources, a high dependence on short-lived, internationally funded programmes, and the rotation of staff. These factors have been linked to a vacillating political will to invest resources over time in policies for science education that could have a cumulative impact on access and quality. Consistency and persistence in up-grading teacher competencies, retaining those who are trained as science teachers in the school system, and husbanding physical resources, so that over time stock accumulates, all seem essential ingredients of sustained improvement.
- The seduction of novelty in curriculum development at the expense of feasible and realistically based curriculum innovation designed to address the needs of the mass of students and teachers working in resource-poor environments. Too often, problems of poor levels of achievement and low-quality access

have provoked curricula change, which has been difficult to implement in the most adaptable parts of the school system (where qualified teachers are plentiful and resources well provided). In less favoured school environments what is proposed may simply not be feasible. Thus, curricula which stress frequent practical activity in small groups in well-funded laboratory environments may simply increase the difficulties of teaching science effectively to the majority. Curriculum development based on thorough appreciation of what is the case, rather than what it ought to be in terms of school conditions, would seem to have more chance of effective implementation.

- A lack of co-ordination between the various actors involved in science education provision, whether at the central, regional or institutional level. As argued above, all too often curriculum developers do not link up with planners and with science inspectors and officers in the Secondary Education Division. Likewise, examination boards and teacher trainers, working at the university level, tend to work independently and do not take into account some of the ministries' concerns and policies. The same can be said at the school level.

Problems of adoption and implementation of more effective science education policies, which are reflected in real changes at the school level, are intimately connected to the existence of effective demand. If pupils, parents and teachers see no particular advantage in accepting costs for the acquisition of new skills, adopting new working practices, and learning new content, they will not be enthusiastic agents of change. This focuses attention on the importance of collateral support for interventions – those things that make it more attractive to change than to stay the same – and on the importance of 'pull' models of innovation rather than 'push' approaches. Changed pedagogy and improved learning outcomes are most likely when science teachers have working conditions that satisfy minimum levels of well-being that allow a focus on their jobs (adequate housing and income), ready access to advice and support from school-based structures and appropriate printed material, and where advisory systems exist. Unless science studies are promoted by society, attractive career pathways are made available, and selection and assessment systems provide incentives for the acquisition of scientific thinking skills, it is unlikely that advocacy or directive approaches alone will popularize the study of science-based subjects. If schools do not place a particularly high value on science subjects, and teachers remain comfortable with what they regard as tried and tested methods, effective demand for change will be weak and there will be little 'pull'.

If established problems in the provision of science education were easy to solve, they would no longer trouble the policy and planning community. Arguably, seeking a solution has become more rather than less urgent. This is the case everywhere, even if circumstances differ.

Some countries continue to grapple with the problems of persistent rural poverty and marginalized groups whose ignorance impedes development. Much of what needs to be applied to ameliorate poverty, the diseases related to it, and the technologies associated with the provision of other basic needs, is known. Whether it is known by those who have the greatest need to apply it remains an open question and the research suggests that this is widely problematic. Basic understanding of science and technology will be a necessary, albeit insufficient, condition for overcoming many of the rural and urban development problems.

Other economies are well advanced in the process of global integration and have, or aspire to, export-led growth, much of which may depend on science-based innovation. Any ambitions to consolidate and improve both international competitiveness and the quality of science-based services have to be fuelled with an improved supply of qualified science-based school leavers until the absorptive capacity of the labour market is satisfied. Those without science-based human resources will find themselves at a disadvantage.

Finally, evidence is accumulating that levels of science achievement in expanded school systems are widely regarded as unsatisfactory, and that learning outcomes may be increasingly dislocated from knowledge and skills of use in the world of work. Pressures arising from selection and examinations can create the kinds of qualification escalation that make it more important to learn to get a job than to learn to do a job. Science education is not immune to these pressures and associated distortions of learning outcomes. But it is the one area of the curriculum (along with its technological applications), where students are invited to understand and manipulate the environment for their own good. It offers access to thinking skills which can help transform both living conditions and the economic basis of production. It can also provide insights into the consequences of adopting different technologies that may have short-term benefits but long-term dangers. For all these reasons, and because the challenges of development are becoming more acute in an increasingly integrated global environment, effective science education provision remains, along with the study of languages and mathematics, vital to these countries' education and development policies.

Solutions to problems of policy and practice are bound to the national systems in which they arise. There is therefore no one solution to what should be done to improve science education. Many things have been identified that can be tried, evidence has been accumulated on promising interventions and the conditions under which they may have the most chance of success, and specific strategies have been isolated that may be developed by planners to monitor and improve access and achievement. Now these have to be interpreted at the country level and below within a historically informed framework, tuned to important contextual realities, and where future effective demand and national development needs for science-based human resources are taken into account.

Bibliography and references

Adamu, A.U. 1992. "Operation, efficiency and desirability of special science schools at secondary level: the Nigerian experience". IIEP research and studies programme. The development of human resources: the provision of science education in secondary schools. Paris: UNESCO/IIEP.

Adey, P.; Shayer, M. 1994. "Really raising standards: cognitive intervention and academic achievement", Routledge.

Allsop, T. 1991. "Practical science in low-income countries". In Woolnough, B., *Practical Science*, Open University Press.

Arghiri, E. 1980. "Technologie appropriée ou technologie sous-développée?" Paris: PUF.

Ausubel, D. 1968. "Educational psychology: a cognitive view". Holt Rhinehart and Winston.

Avalos, B. 1995. "Issues in science teacher education". IIEP research and studies programme. The development of human resources: the provision of science education in secondary schools. Paris: UNESCO/IIEP.

Bishop, J. 1989. "Is the test score decline responsible for the productivity growth declines?" *The American Economic Review*, Vol. 79, No. 1, March, pp 178-197.

Black, P. 1990. "The purposes and roles of assessment in science and technology education". In Layton, D. (ed). *Innovations in science and technology education*, Vol. 3, pp. 15-34. Paris: UNESCO.

Black, H.; Dockrell, W. 1984. "Criterion referenced assessment in the classroom", Scottish Council for Educational Research.

Bude, U.; Lewin, K. (eds). 1996. "Science and agriculture in primary school leaving examinations in Eastern and Southern Africa: constructing tests, analysing results and improving assessment quality". Bonn: Deutsche Stiftung für Internationale Entwicklung.

Caillods, F.; Göttelmann-Duret, G., Radi, M. and Hddigui El, M. 1997. "La formation scientifique au Maroc". Paris:UNESCO/IIEP.

Caillods, F. 1994. "Converging trends amidst diversity in vocational training systems". In *International Labour Review*, Vol. 133, No. 2. Geneva: ILO.

Caillods, F.; Göttelmann-Duret, G. 1991. "Science provision in academic secondary schools, organisation and condition". IIEP Mimeo. Paris: UNESCO/IIEP.

Calzadilla, V.; Bruni Celli, J. 1994. "La educación técnica media en Venezuela". Caracas: Publicaciones Cinterplan.

Cheng, Kai-ming. 1991. "Conformity and adaptability. Explaining educational practices in East Asia". Paper prepared for the Ninth International Forum, Harvard Graduate School of Education. 12-13 April.

Chew, D. 1992. "Civil service pay in South Asia". Geneva:ILO.

Centre d'Etudes et de Recherches sur les Qualifications. 1991. "Les emplois et l'insertion professionnelle des diplômés de l'enseignement supérieur". In *Documents de travail*, CEREQ, N°. 64, Avril.

Centre d'Etude des Revenus et des Coûts (CERC). 1990. "Les rémunérations des jeunes à l'entrée dans la vie active", N° 99, 4ème trimestre. Dans *La documentation française*, Paris.

Deleage, J.-P.; Souchon, C. 1993. L'éducation pour l'environnement et son insertion dans l'enseignement secondaire. Programme de recherche et d'études de l'IIPE. Développement des ressources humaines : planifier l'offre d'enseignement scientifique à l'école secondaire. Paris: UNESCO/IIPE.

De Moura Castro, C.; Feonova, M.; Litman, A. 1997. "Education and production in Russia: what are the lessons?" Paris: UNESCO/IIEP.

De Moura Castro, C. 1995. "Training policies for the end of the century". IIEP Research and Studies Programme. The development of human resources. New trends in technical and vocational education. Paris: UNESCO/IIEP.

Denison, E. 1962. "Sources of economic growth in the United States". Washington: Committee for Economic Development.

Dore, R.P. 1989. "Latecomers' problems". In *European Journal of Development Research*, Vol. 1, No. 1. London: Frank Cass and Co.Ltd.

Dore, R.P. 1976. "The diploma disease: education, qualification and development". London: Allen and Unwin.

Driver, R. 1983. "The pupil as scientist?" Milton Keynes (United Kingdom):The Open University Press.

Duru-Bellat, M. 1994. "Filles et sciences" : rapport final. Université de Bourgogne, Dijon, France.

Eisemon, T.O. 1992. "Language issues in scientific training and research in developing countries." PHREE background paper series; PHREE/92/047. Washington, D.C.: The World Bank.

Fransman, M.; King, K. (Eds.) 1984. "Technological capability in the Third World". London: Macmillan.

Freeman, C. 1989. " New technology and catching up". In *The European Journal of Development Research,* Vol. 1, No. 1, June.

Frith, D.; Mackintosh, H.A. 1984. "Teacher's guide to assessment", Stanley Thornes.

Fuller, B.; Hue, H.; Snyder, C.W. 1994. "When girls learn more than boys: the influence of time in school and pedagogy in Botswana". In *Comparative Education Review*, Vol. 38, No. 3.

Furtado, C. 1976. "Le Mythe du développement économique". Paris: Anthropos. 2e ed. 1984, p.111.

Gagne, R. 1965. "The conditions of learning". Holt Rhinehart and Winston.

Gallart, M.A. 1990. "Technical and technological education in Argentina". IIEP Mimeo. Paris: UNESCO/IIEP.

Garnier, M.; Hage, J. 1994. "Type de capital humain et croissance économique". *Les dossiers d'éducation et formations*. N° 42. Paris: Ministère de l'Education nationale; Direction de l'Evaluation et de la Prospective. Décembre.

George, J.; Glasgow, J. 1988. "Street science and conventional science in the West Indies". In *Studies in Science Education 1*, pp 109-118.

Giordan, A., Girault, Y. 1994. "Les aspects qualitatifs de l'enseignement des sciences dans les pays Francophones". Programmes de recherche et d'études de l'IIPE. Développement des ressources humaines : planifier l'offre d'enseignement scientifique à l'école secondaire. Paris: UNESCO/IIPE.

Government of Malaysia. 1990. Malaysia Human Resources Development Plan Project, Report on Education.

Government of Malaysia. 1985. "Education in Malaysia". EPRD, Ministry of Education.

Greffe, X. 1997. "La mise en place de formations initiales en alternance: enjeux, problèmes et solutions". IIEP research and studies programme. The development of human resources. New trends in technical and vocational education. Paris: UNESCO/IIEP.

Haddad, W.D. 1995. "Education policy-planning process: an applied framework". Fundamentals of Educational Planning series. No. 51. Paris: UNESCO/IIEP.

Haddad, W.D.; Za'rour, G.I. 1986. "Role and educational effects of practical activities in science education". Discussion Paper: *Education and Training Series*, EDT 51. Washington, D.C.: World Bank.

Hage, J.; Garnier, M. 1992. "Strong States and educational expansion: France versus Italy in the political construction of education". *The State, School Expansion, and Economic Change*, pp. 155-172, New York: Praeger.

Hage, J.; Garnier, M.; Fuller, B. 1988. "The active State, investment in human capital and economic growth: France 1825-1975". In *American Sociological Review*, Vol. 53, December, pp. 824-837.

Harding, J. 1992. "Breaking the barrier: girls in science education". IIEP research and studies programme. The development of human resources: the provision of science education in secondary schools. Paris:UNESCO/IIEP.

Harvey Brooks. 1988. "Scientific literacy and the future labor force". Mimeo.

Head, J. 1985. "The personal response to science", Cambridge Education in Science Series.

Herrera, A. 1978. "Technolgias cientificas y tradicionales en los paises en desarollo" In *Ressources naturelles, technologie et indépendance*. Mexico: Commercio Exterior, Vol.28, No. 12, December.

Hicks, G.L.; Redding, S.G. 1983. "The story of the East Asian economic miracle." In *Euro-Asia Business Review* 2 (3) and 2 (4).

Hofstein, A.; Lunetta, V.N. 1982. "The role of the laboratory in science teaching: neglected aspects of research". In *Review of Educational Research*, Vol. 52, No. 2, pp. 201-217.

IEA. 1991 and 1992. International Association for the Evaluation of Educational Achievement. Science achievement in twenty-three countries. The second IEA Science Study (SISS); (for the first IEA Science Study, see Vol.3). (See individual entries in this bibliography under: the IEA study of Science I: (Vol. I) Rosier, M.J.; Keeves, J.P. 1991 (Eds.); The IEA: study of Science II: (Vol. 2) Postlethwaite, T.N.; Wiley, D.E. 1992.; The IEA study of Science III: (Vol. 3) Keeves, J.P.1992 (Ed.). Oxford (United Kingdom): Pergamon Press.

IEAP/ETS. 1992a and 1992b. International Assessment of Educational Progress, Educational Testing Service: Princeton (N.J.): ETS. (See Lapointe et al., 1992a and 1992b).

IIEP. 1995. South African Consortium for Monitoring Educational Equality (SACMEQ) IIEP Newsletter. Paris: UNESCO/IIEP.

IIEP. 1984. Educational planning in the context of current development problems. Papers presented at an IIEP Seminar, 3-8 October, 1993, Vols. 1 & 2, October. Paris: UNESCO/IIEP.

ILO. 1981. The paper qualification syndrome and the unemployment of school leavers: a comparative sub-regional study of four West African and four East African countries, 2 Vols. Geneva:ILO.

Jong Ha Han. 1993. "The role of education in Korean industrialization: an emphasis on human resource development". Paper presented for the international symposium of the economics of education. Manchester: British Council.

Katz, J. 1984. "Technological innovation, industrial organization and comparative advantages of Latin American metalworking industries". In Fransman and King (Eds.) "Technological capability in the Third World". London: MacMillan.

Kearns, D.T.; Doyle, D.P. 1988. "Winning the brain race: a bold plan to make our schools competitive". San Francisco (CA): ICS Press.

Keeves, J. 1994. "National examinations: design, procedures and reporting". *Fundamentals of Educational Planning series, No. 50*. Paris: UNESCO/IIEP.

Keeves, J. (Ed.). 1991. The IEA Study of Science III: changes in Science education and achievement: 1970 to 1984, Vol.3. Oxford: Pergamon Press.

Kellaghan, T.; Greaney, V. 1992. "Using examinations to improve education: a study in fourteen African countries". World Bank technical papers, Africa Technical Department, No. 165. Washington, D.C.: The World Bank.

Keller, E. 1986. "Reflections on gender and science", Yale University Press.

King, K.; Layton, D.; Young, B. 1989. "Educating for capability: the role of science and technology education". London: British Council.

King, K. 1982. "Science, technology and education research in India". A discussion paper. Centre for African Studies. University of Edinburgh.

Knamiller, G.W. 1988. "Linking traditional science with school science in Malawi". Mimeo. University of Leeds.

Lall, S. 1994. "Technological capabilities in the uncertain quest". In Salomon, J. (Ed.) "The uncertain quest: science, technology and development". Tokyo; New York; Paris: United Nations University Press.

Lapointe, A.E.; Mead, N.A.; Askew, J.M.1992a. "Learning mathematics".The International Assessment of Educational Progress (IAEP), Center for the Assessment of Educational Progress at Educational Testing Service, Princeton, USA.

Lapointe, A.E.; Mead, N.A.; Askew, J.M.1992b. "Learning science 1992". The International Assessment of Educational Progress (IAEP), Center for the Assessment of Educational Progress at Educational Testing Service, Princeton, USA.

Layton, D. (Ed). 1994. *Innovations in science and technology education*, Vol.5. Paris: UNESCO.

Layton, D. 1993. "Technology's challenge to science education: cathedral, quarry or company store?" Buckingham (United Kingdom) and Bristol (USA): Open University Press.

Leite, E.M.; Caillods, F. 1987. "Education, training and employment in small-scale enterprises: three industries in Sao Paulo, Brazil". Paris: UNESCO/IIEP.

Lewin, K. 1995. "Development policy and science education in South Africa: reflections on post-Fordism and Praxis". In *Comparative Education*, Vol. 31, No. 2, pp. 203-222.

Lewin, K. 1994. "Education and development, the issues and the evidence". ODA Monograph No. 6, London.

Lewin, K. 1992. "Science education in developing countries: issues and perspectives for planners". IIEP research and studies programme. The development of human resources: the provision of science educaion in secondary schools. Paris: UNESCO/IIEP.

Little et al. 1987. Why do students learn? A six country study of student motivation. IDS research report.

Lunetta, V.; Hofstein, A.(Eds.). 1990. "Simulation and laboratory practical activity". In Woolnough, B., *Practical Science,* pp. 125-137. Milton Keynes (United Kingdom): Open University Press.

Martinand, J.-L. 1994. "La technologie dans l'enseignement général : les enjeux de la conception et de la mise en oeuvre". Programme de recherche et d'études de l'IIPE. Développement des ressources humaines : planifier l'offre d'enseignement scientifique à l'école secondaire. Paris. UNESCO/IIPE.

Middleton, J.; Ziderman, A.; Adams, A.V. 1993. "Skills for productivity in vocational education and training in developing countries". Washington, D.C.:The World Bank. New York (N.Y.); Oxford University Press.

Middleton, J.; Ziderman, A.; Adams, A.V. 1991. "Vocational and technical education and training". Washington, D.C.: The World Bank.

National Science Foundation. 1993. "Human resources for science and technology". The Asian region surveys of science resources series.

Nielsen, H.D.; Tatto, M.; Djalil, A.; Kularatne, N. 1991. "The cost-effectiveness of distance education for teacher training". Bridges research report series: 009. Cambridge (Mass.): Harvard University.

OECD-CEREQ. 1994. Seminar on apprenticeship, alternance and dual systems: dead-ends or highways to the future. Marseilles, France, April. Paris: OECD.

OECD. 1993. "From higher education to employment", Synthesis Report. Paris: OECD.

OECD. 1992. "Technology and the economy: the big relationship". Paris: OECD.

Oxenham, J.C.P. (Ed). 1984. "Education versus qualifications". A study of relationships between education, selection for employment and productivity of labour. London, United Kingdom: George Allen and Unwin.

Pennycuick, D. 1990. "Factors influencing the introduction of continuous assessment systems in developing countries". In Layton, D. (Ed). *Innovations in Science and Technology Education,* Vol. 3. Paris: UNESCO.

Postlethwaite, T.N.; Wiley, D.E. 1992. The IEA Study of Science II: Science achievement in twenty-three countries,Vol. 2. Oxford, New York, Seoul, Tokyo: Pergamon Press.

Prais, S.J.; Wagner, K. 1986. "Schooling standards in England and Germany: some summary comparisons bearing on economic performance". *Compare,* Vol. 16, No.1.

Robinson, D. 1990. "Civil service pay in Africa". Geneva:ILO.

Robson, M. 1992. "Introducing technology through Science Education: a case study from Zimbabwe". In *Science, Technology and Development*, Vol. 10, No. 2.

Rosier, M.J.; Keeves, J.P. 1991 (Eds.). The IEA Study of Science I: Science education and curricula in twenty-three countries, Vol 1. Oxford: Pergamon Press.

Ross, K.N. (1994a). "The construction of test blueprints". IIEP training materials (mimeo). Paris: UNESCO/IIEP.

Ross, K.N. (1994b). "Draft plan for a national survey of the basic reading achievement of primary schools in the Grade Six level". IIEP training materials (mimeo). Paris: UNESCO/IIEP.

Ross, A.; Lewin, K. 1992. "Science kits in developing countries: an appraisal of potential". IIEP research and studies programme. The development of human resources: the provision of science education in secondary schools. Paris: UNESCO/IIEP.

Sagasti, F. 1989. "Science and technology policy research for development: an overview and some policies from a Latin American perspective". *Bulletin of Science, Technology and Society 9,* No. 1.

Salomon, J-J. 1984. "La science ne garantit pas le développement". In *Futuribles*, No. 78, juin.

Salomon, J.-J.; Lebeau, A. 1993. "Mirages of development". Boulder, Colo.: Lynne Rienner. Originally published in French in 1988 as *L'ecrivain public et l'ordinateur*. Paris: Hachette.

Salomon, J-J.; Sagasti, F.; Sachs-Jeantet, C. 1994. "The uncertain quest: science, technology and development". Tokyo; New York; Paris: United Nations University Press.

Sharifah Maimunah; Lewin, K. (Eds). 1993. "Insights into science education: planning and policy priorities in Malaysia". IIEP research and studies programme. The development of human resources: the provision of science education in secondary schools. Paris: UNESCO/IIEP; Malaysia, Ministry of Education.

Schiefelbein, E.; Farrell, J.P.; Sepulveda-Stuardo, M.A. 1983. The influence of school resources in Chile: their effect on educational achievement and occupational attainment. World Bank staff working papers. Washington, D.C.: The World Bank.

Schultz, T.W. 1961. "Investment in human capital". *American Economic Review* (51).

Shayer; Adey. 1981. "Towards a science of science teaching: cognitive development and curriculum demand". Heinemann Education.

Stuart, J. 1991. Classroom action research in Africa; a Lesotho case study of curriculum and professional development. In Lewin, K. with Stuart, J.S. (Eds.). "Educational innovation in developing countries: case studies of changemakers". London: MacMillan.

Swift, D. 1983. "Physics for rural development: a source book for teachers and extension workers in developing countries". Chichester: John Wiley.

Tan, J.-P.; Mingat, A. 1992. Education in Asia - a comparative study of cost and financing in World Bank regional and sectoral studies, Washington, D.C.: The World Bank.

Tatto, M.T.; Nielsen, H.; Cummings, W.; Kularatne, N.; Dharmdasaln, Kh. 1992. "Educating primary school teachers: comparing the effects and costs of different approaches to train teachers". In *The Forum*, Vol. 2, Issue 1, October.

UNESCO. 1985. Low cost equipment for science and technology education. UNESCO Document No. ED/85/WS/60. Paris: UNESCO.

Wad, A. 1994. "Science and technology policy." In *The uncertain question: Science, technology, and development*, Salomon, J.-J.; Sagasti, F.R.; Sachs-Jeantet; C. (Eds.). Tokyo, New York, Paris: The United Nations University Press.

Wade, R. 1989. "Economic theory and the role of government in East Asian industrialization". New Jersey, USA: Princeton University Press.

Walberg, H. 1991. "Improving school science in advanced and developing countries". In *Review of Educational Research*, Vol. 61(1), pp. 25-69.

Walters, P.B.; Rubinson, R. 1983. "Educational expansion and economic output in the United States, 1890-1969: a production function analysis". *American Sociological Review*, Vol. 48 (August: 480-493).

Wanjala Kerre, B. 1994. "Technology education in Africa". In Layton, D.(Ed.). *Innovations in science and technology education*, Vol.5. Paris: UNESCO.

Ware, S. 1992a. "Secondary school science in developing countries: status and issues". Education and Employment Division, Population and Human Resources Department, PHREE Background Paper Series, Document No. PHREE/92/53. Washington, D.C.: World Bank.

Ware, S. 1992b. "The education of secondary science teachers in developing countries". Education and Employment Division, Population and Human Resources Department. PHREE Background Paper Series, Document No. PHREE/92/68. Washington, D.C.: World Bank.

Weizsäcker, E.U. von; Swaminathan, M.S.; Lemma, A. 1983. "Endogenous generation of technology instead of imitated innovation". New frontiers in technology application: integration of emerging and traditional technologies. In *Science and technology for development* series, Vol. 2. Dublin: Tycooly.

Welford, G. 1990. "Assessment of practical work in school science". In Layton, D., (Ed.), *Innovations inScience and Technology Education*, Vol. 3, pp. 37-54. Paris: UNESCO.

White, R. 1991. "Episodes and the purpose and conduct of practical work". In Woolnough, B., "Practical Science", Open University Press.

Wilson, B. 1981. "Cultural contexts of science and mathematics education: a bibliographic guide". Centre for Studies in Science Education, University of Leeds, United Kingdom.

Wolf, A. 1993. "Assessment and problems in a criterion-based system". London: Further Education Unit Occasional Paper, Department for Education, Her Majesty's Stationery Office (HMSO), London.

World Bank. 1993. "The East-Asian miracle, economic growth and public policy". World Bank: Washington, D.C..

Studies prepared within the framework of the International Institute for Educational Planning (IIEP) Project on "Planning science education provision in general secondary schools", directed by Françoise Caillods with Gabriele Göttelmann-Duret

Adamu, A.U. 1992. "Operation, efficiency and desirability of special science schools at secondary level: the Nigerian experience". IIEP research and studies programme. The development of human resources: the provision of science education in secondary schools. Paris: UNESCO/IIEP.

Avalos, B. 1995. "Issues in science teacher education". IIEP research and studies programme. The development of human resources: the provision of science education in secondary schools. Paris: UNESCO/IIEP.

Caillods, F.; Göttelmann-Duret, G., Radi, M. and Hddigui El, M. 1997. "La formation scientifique au Maroc". Paris:UNESCO/IIEP.

Deleage, J. P.; Souchon, C. 1993. L'éducation pour l'environnement et son insertion dans l'enseignement secondaire. Programme de recherche et d'études de l'IIPE. Développement des ressources humaines : planifier l'offre d'enseignement scientifique à l'école secondaire. Paris: UNESCO/IIPE.

Giordan, A., Girault, Y. 1994. "Les aspects qualitatifs de l'enseignement des sciences dans les pays francophones". Programmes de recherche et d'études de l'IIPE. Développement des ressources humaines : planifier l'offre d'enseignement scientifique à l'école secondaire. Paris: UNESCO/IIPE.

Harding, J. 1992. "Breaking the barrier: girls in science education". IIEP research and studies programme. The development of human resources: the provision of science education in secondary schools. Paris:UNESCO/IIEP.

Lewin, K. 1992. "Science education in developing countries: issues and perspectives for planners". IIEP research and studies programme. The development of human resources: the provision of science educaion in secondary schools. Paris: UNESCO/IIEP.

Martinand, J.-L. 1994. "La technologie dans l'enseignement général : les enjeux de la conception et de la mise en oeuvre". Programme de recherche et d'études de l'IIPE. Développement des ressources humaines : planifier l'offre d'enseignement scientifique à l'école secondaire. Paris. UNESCO/IIPE.

Ross, A.; Lewin, K. 1992. "Science kits in developing countries: an appraisal of potential". IIEP research and studies programme. The development of human resources: the provision of science education in secondary schools. Paris: UNESCO/IIEP.

Sharifah Maimunah; Lewin, K. (Eds). 1993. "Insights into science education: planning and policy priorities in Malaysia". IIEP research and studies programme. The development of human resources: the provision of science education in secondary schools. Paris: UNESCO/IIEP; Malaysia, Ministry of Education.

Unpublished studies:

Caillods, F.; Göttelmann-Duret, G. 1991. "Science provision in academic secondary schools, organisation and condition". IIEP Mimeo. Paris: UNESCO/IIEP.

Daboué, J. 1990. Burkina Faso : "L'état de la formation scientifique dans l'enseignement secondaire général". Ministère des enseignements secondaires, supérieur et de la recherche scientifique. Direction des études et de la planification.

George, T.C. 1990. "The condition of science provision in academic secondary education in Papua New Guinea". IIEP Mimeo. Paris: UNESCO/IIEP.

Iwasaki, K. 1991. "The condition of science provision in academic secondary education in Japan", IIEP Mimeo. Paris: UNESCO/IIEP.

Kann, U; Nganunu, M. 1991. "The condition of science provision in academic secondary education in Botswana". IIEP Mimeo. Paris: UNESCO/IIEP.

Lee, W. K. 1991. "The condition of science provision in academic secondary education in the Republic of Korea". IIEP Mimeo. Paris:UNESCO/IIEP.

Lemos, M.L. 1990. Argentina :"Situación de la formación cientifica en la enseñanza secundaria general". IIEP Mimeo. Paris: UNESCO/IIEP.

Longo, T.M. 1990. "La formation scientifique dans les écoles italiennes". IIEP Mimeo. Paris: UNESCO/IIEP.

Obura, A. 1992. "The condition of science provision in academic secondary education in Kenya", IIEP Mimeo. Paris: UNESCO/IIEP.

Radi, M. 1991. "L'état de la formation scientifique dans l'enseignement secondaire général au Maroc". IIEP Mimeo. Paris: UNESCO/IIEP.

Rojo, F.; Martinez, R. 1991. "The condition of science provision in academic education in Mexico". IIEP Mimeo. Paris:UNESCO/IIEP.

Seck Fall, S. 1991. "L'état de la formation scientifique dans l'enseignement secondaire général au Sénégal". IIEP Mimeo. Paris: UNESCO/IIPE.

Schakmann, L. von; Zepeda, S.; Toro, E. 1992. "Encuesta internacional sobre la situacion de la formacion cientifica en la enseñanza secundaria en Chile". IIEP Mimeo. Paris: UNESCO/IIEP.

Sharifah Maimunah. 1991. "The condition of science provision in academic secondary education in Malaysia". IIEP Mimeo. Paris: UNESCO/IIEP.

Sweilem, M.A.; Aleweher, M.; Mdanat, H.; Zu'bu, T. 1991. "The condition of science provision in academic secondary education in Jordan". IIEP Mimeo. Paris: UNESCO/IIEP.

Walsh, M.; Gregorio, L.; Maclean, R. 1994. "The training of secondary science teachers". IIEP Mimeo. Paris: UNESCO/IIEP.

Wilailak, W. 1991. Ministry of Education. "The condition of science provision in academic secondary education in Thailand". IIEP Mimeo. Paris: UNESCO/IIEP.

IIEP publications and documents

More than 1,120 titles on all aspects of educational planning have been published by the International Institute for Educational Planning. A comprehensive catalogue, giving details of their availability, includes research reports, case studies, seminar documents, training materials, occasional papers and reference books in the following subject categories:

Economics of education, costs and financing.

Manpower and employment.

Demographic studies.

The location of schools (school map) and sub-national planning.

Administration and management.

Curriculum development and evaluation.

Educational technology.

Primary, secondary and higher education.

Vocational and technical education.

Non-formal, out-of-school, adult and rural education.

Disadvantaged groups.

Copies of the catalogue may be obtained from the IIEP Publications Unit on request.